Collection "Guías imprescindibles de flora, 1"

Wild Flowers of Ordesa and Monte Perdido National Park

(Spanish Pyrenees)

José Luis Benito Alonso

Jolube Consultor Botánico y Editor

Jaca, 2014

CONTENTS

What this book contains

This guide to the flowers of Ordesa and Monte Perdido National Park —a World Heritage Site— can be considered the first introduction to the flora of this outstanding natural reserve which harbours —the National Park and its peripheral area altogether— more than 1,400 plants, 146 of which have been treated in this guide. If you would like to enlarge your information on this subject, you will find a concise bibliography at the end of this book.

Our aim is to bring the reader some of the most striking and distinctive plants that exist in this National Park by using a clear, accessible language. The species are arranged in different environments (forests and scrubs, clearings and megaphorbs, pastures, meadows, humid sites, rocks, screes and high mountains), although they might also be found in other plant communities. Representative plants in each ecosystem have been chosen, both common and rare, the latter having their prime and sometimes single populations within the Park, so they endow our flora with singularity.

The plants are classified alphabetically within each habitat according to their scientific name. Each species is illustrated with a large image and very often with others so as to enhance various features. The plant profile also includes the Latin name, along with that of the botanical family it belongs to; as a complementary information synonyms are added in small print, followed by a selection of vernacular names, the English ones in red type, **French ones in blue type**, the Spanish ones in rounded characters and the *Aragonese ones in italic type*. Not all the species have a vernacular name while others just have one among the three languages. An index can be consulted at the end of the book.

Afterwards, the occurrence of each plant in different sectors of the Park is stated, ordered from west to east, by using the following codification: **B**: Bujaruelo valley; **T**: Torla and its surroundings; **O**: Ordesa valley; **V**: Vió valley; A: Añisclo valley; **E**: Escuaín valley; **P**: Pineta valley; **C**: Chisagüés valley. Later on, comes the altitudinal range where the species has been recorded (in brackets for an extreme or unusual altitude). Next is a series of abbreviations that provide some details about the species growth form and its geographical range (chorology) as well.

Plant life-forms

This is a classification regarding the way plants spend the unfavourable season.

Ch: Chamaephyte. Small shrub.
Ep: Epiphyte. Climbing plant.
G: Geophyte. Bulbous or rhizomatous plant.
H: Hemicryptophyte. Plant whose leaves grow at ground level.
MP: Megaphanerophyte. Tree.
NP: Nanophanerophyte. Bush or shrub.
P: Phanerophyte. Tree or bush.
Th: Therophyte. Annual herb, with a short vital cycle.
dec.: Deciduous, losing its leaves annually.
everg.: Evergreen, having green leaves throughout the year.

Distribution

In many cases, the plants which occur in the Park can also be found in other parts of the world. Scientists have divided our planet into different regions depending on their climatic and ecological affinities.

Alp.: Alpine, from the alpine-type mountains in Europe.
Arct.: Arctic.
Atl.: Atlantic, from the areas near the Atlantic ocean.
Bor.-alp.: Boreal-alpine, from the Boreal zone and the alpine-type mountains.
Ceven.: Cevennian, from the Cevennes mountains in France.
Circumb.: Circumboreal, from all over the Northern hemisphere.
Eur.: Euro-Siberian, from central and east Europe.
Iber.-Pen.: from the Iberian Peninsula.
Late: prefix meaning "in the broad sense", that is to say it goes well beyond the region considered.
Med.: Mediterranean, from the coastal and inland areas that surround the Mediterranean Sea.
Mount.: from the mountains.
Pyr.: occurring exclusively in the Pyrenees.
Pyr.-Cant.: occurring exclusively in the Pyrenees and the Cantabrian Mountains.
Plurireg.: Pluri-regional, that occurs in many regions on the Earth.
Submed.: Sub-mediterranean, that is the transitional zone between Mediterranean and central European climates, in this case mainly located in the Pre-Pyrenees and adjacent zones climatically similar.
C: Central. **E**: East. **N**: North. **S**: South. **W**: West.

Finally the abundance of each plant within this territory is assessed by means of 7 categories: very rare **(RR)**, having less than four locations; rare **(R)**; scarce **(E)**; frequent **(F)**; common **(C)**; very common **(CC)**; and general **(CCC)** for those landscape-forming species.

Translation from Spanish by José Vicente Ferrández Palacio. Special thanks to Amanda Tyson for her help in improving the final English version of this book.

Introduction

Ordesa and Monte Perdido National Park (PNOMP in its Spanish acronym) is located in the central Pyrenees of Huesca (Aragón, Spain), in northwestern Sobrarbe and bordering the French Pryrenees National Park to the north. It was founded in 1912, when 2,200 hectares corresponding to the forests of Ordesa valley were protected. In 1982 its surface area was enlarged until the current 15,608 hectares that include four valleys besides Monte Perdido, the highest calcareous massif in Europe. From west to east we find: Ordesa (Arazas river), Añisclo (Bellós river), Escuaín (Yaga river) and the uppermost Pineta valley (Cinca river), all of them located within the Cinca basin. The peripheral area surrounding the Park is made up of Bujaruelo valley (upper Ara river), the southern mountainside of Vió valley, the right bank of Airés ravine in Puértolas-Bestué valley, the middle section of Pineta valley and finally the entire La Larri basin up to La Munia Peaks.

Ordesa and Monte Perdido was the second National Park to be created in Spain, after Covadonga. It was also declared a **Biosphere Reserve-MAB** by the UNESCO in 1977, a **World Heritage Site** in 1997, and a **Site of Communitary Importance** by the European Union in 2004.

Among the Spanish National Parks, PNOMP has the widest altitudinal range, 2,655 metres, its lowest point being at 700 metres in Añisclo, while its highest point, Monte Perdido, rises to 3,355 metres, the third highest elevation in the Pyrenees.

The climate of this territory is ruled by two major features: on one hand, an equinoctial rainfall pattern characterized by the highest amount registered in autumn and spring, while the lowest is had in winter, with great irregularities throughout the years; on the other hand, a marked continentality because of the distance from the seas, along with important mountains that obstruct the progress of fronts. Some areas of well-marked Mediterranean character, such as Añisclo, co-exist with others that are more humid and remind us of Central Europe; and finally, we have the alpine nature on the summits.

All this means that practically a sample of every plant community occurring on limestone in the southern side of central Pyrenees can be found within the Park, ranging from lowland Mediterranean to high-mountain alpine types. This wide altitudinal range allows different sorts of vegetation to be distinguished, corresponding to different geobotanical belts or zones.

The Mediterranean and continental sub-mediterranean zones appear in the most sheltered parts of Añisclo and Escuaín gorges, with just a few winter frosts and a certain summer drought, where holm oak and *Quercus subpyrenaica* woods occur. The continental montane zone, ranging between 1,000 and 1,700 metres is quite cool, with temperature inversion and strong contrasts from south to north-facing slopes, and contains montane forests such as pine, beech, silver fir and mixed woods. The Mediterranean mountain zone (oro-Mediterranean) is peculiar to sunny slopes up to 2,200 metres, where summer low atmospheric humidity combines with high mountain cold, and gives shelter to pinewoods, hedgehog heaths (spiny cushion-like scrubs) and open stony pastures. Finally we find the treeless alpine zone, which is cold, sunny in winter and cool in summer, allowing just a short growth period, with some areas of perpetual snow.

FRANCIA
Valle de Bujaruelo
bco. Lapazosa
bco. Otal
Gavarnie
Estaubé
Astazu Or.
3071
Tucarroya
2819
Pineta
2861
Astazu Occ.
3015
Brecha de Rolando
2807
Taillón
3146
Gabietos
3036 - 3034
El Casco
3012
La Torre
3015
Picos de la Cascada
Pico Blanco
2919
Balcón de Pineta
río Ara
Cdo. del Descargador
Cdo. de la Catuerta
Mondarruego
2847
Cdo. de Millaris
Faja Luenga
Espalda
3075
I. Helado
Monte Perdido
3355
Gallinero
2752
Descargador
2627
Ordesa
Escaleras
3027
Tozal del Mallo
2254
Góriz
Carriata
Faja Racón
Cotatuero
Millaris
2619
Arrablo
2792
Cdo. de Góriz
Pte. de Los Navarros
Cola de Caballo
Custodia
2519
Turieto Bajo
Tobacor
2779
río Arazas
Gradas de Soaso
Faja Pelay
Sierra Custodia
Capradiza
Diazas
Punta Acuta
2242
La Carquera
Cuello Gordo
Torla
Sierra de las Cutas
Sarracils
Pueyo de Mondicieto
2384
Añisclo
río Bellós
La Pardina
bco. del Furco
Valle de Vió
Cuello Arenas
Ripareta
Broto
Oto
La Estiba
2004
Selba Plana
bco. de Aso
Buesa
Fanlo
Mondoto
1957
Sarvisé
Buisán
Nerín
Sercué
Portiello
© JOLUBE & SANÉ, 2002
río Ara
bco. del Chate
Valle de Broto
Ceresuela
Vió
Buerba
Valle de La Solana
Asín de Broto
río Yesa
Cájol
Yeba
Semolué
Burgasé
Castellar
Geré
Giral
San Felices

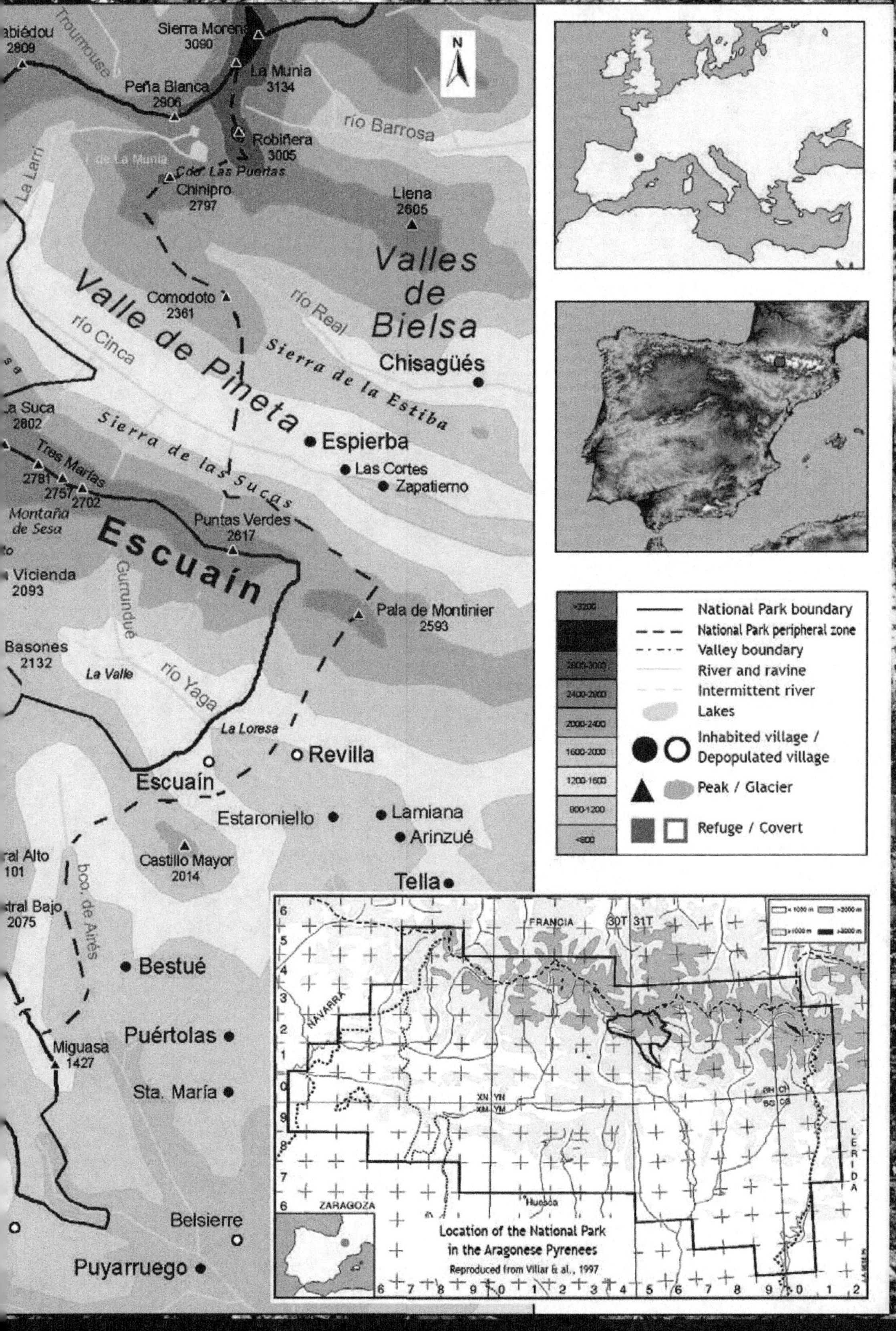

Sierra Morena
3090
La Munia
3134
Peña Blanca
2906
Troumouse
río Barrosa
Robiñera
3005
Col. Las Puertas
Chinipro
2797
Liena
2605
La Larri
Valles de Bielsa
Valle de Pineta
Comodoto
2361
río Real
río Cinca
Sierra de la Estiba
Chisagüés
La Suca
2802
Sierra de las Sucas
Espierba
Las Cortes
Zapatierno
Tres Marías
2781
2757
2702
Montaña de Sesa
Puntas Verdes
2617
Escuaín
Vicienda
2093
Gurrundué
Pala de Montinier
2593
Basones
2132
La Valle
río Yaga
La Loresa
Revilla
Escuaín
Estaroniello
Lamiana
Arinzué
Castillo Mayor
2014
Tella
bco. de Airés
Bestué
Puértolas
Miguasa
1427
Sta. María
Belsierre
Puyarruego
National Park boundary
National Park peripheral zone
Valley boundary
River and ravine
Intermittent river
Lakes
Inhabited village / Depopulated village
Peak / Glacier
Refuge / Covert
>3200
2800-3000
2400-2800
2000-2400
1600-2000
1200-1600
800-1200
<800
FRANCIA
30T
31T
NAVARRA
ZARAGOZA
Huesca
LERIDA
Location of the National Park
in the Aragonese Pyrenees
Reproduced from Villar & al., 1997

FORESTS and SHRUBS

The first impression one gets when entering Ordesa and Monte Perdido is that it is forest-cloaked. However, only 21 % of its territory is actually covered by forests, which form some masses distributed all along the four valleys: Ordesa to the W, Añisclo to the S, Escuaín to the ESE and Pineta to the NE. Practically all the forests typical of the calcareous southern Pyrenees are found within this protected area, with the exception of the Pyrenean black pine woods (*Pinus nigra* subsp. *salzmannii*) that almost reach the Park southernmost limits and the pedunculate oak woods (*Quercus robur*) typical of the Atlantic foothill zone.

Within the PNOMP both beechwoods and beech-silver fir mixed woods are dominant, representing more than one third of all forest masses, with Scots pine woods immediately behind them. Further apart, with just 10%, are the mountain pine subalpine woods (*Pinus uncinata*), with about the same area being occupied by holm oak woods. The rest corresponds to hazel-mixed woods (c. 7%); *Quercus subpyrenaica* woods (4,5%) and willow thickets (> 1%). As for the local distribution of the various types of forests in each valley there are strong dissimilarities because of their different positions.

Ordesa faces West and, along with Añisclo, is the most forested valley. The lowest parts of its southern slopes are covered with extensive Scots pine woods, while most north-facing mountainsides are covered with beechwoods, silver fir woods and mossy pinewoods that catch humidity from the oceanic fronts. However, all three types of forests also reach the south-facing slopes from the middle zone up to Soaso, as the narrowness of the valley causes a blockage of cold air in the bottom and the subsequent condensation of humidity, a phenomenon called "temperature inversion". The main interest of the Ordesa forests lies in the fact that they have not been exploited for 100 years.

Unfortunately **Pineta** valley only gets a little protected area within the Park, despite its strongly forested nature. Two-thirds of the protected forests in Pineta are beechwoods or beech-silver fir mixed woods. From among all four valleys this is the broadest one so river verge communities are well-developed, with the rare Violet Willow, *Salix daphnoides*. Without a doubt, the most remarkable feature is the occurrence of the largest best-preserved mixed wood growing on alluvial terraces in the entire Aragonese Pyrenees.

Escuaín is a relatively short valley where the landscape is greatly intervened by man, so much is this the case that it lacks mountain pine woods. It is also the most continental and sub-mediterranean of all four, as it gives shelter to nearly all *Quercus subpyrenaica* woods in the Park and 65% of its forests are Scots pine woods, while wet forests have little importance. Finally, some dry spurs hold remarkable holm oak thickets on their northernmost European distribution limit.

Añisclo gathers the largest diversity and contrasts in terms of vegetation for various reasons: it is the longest valley -more than 20 km–, and descends zigzagging from North to South; it has got some short transverse valleys as well, and it also displays a wide altitudinal range from 700 to 3,000 metres, containing all of the geobotanical belts in the Park, from the Mediterranean foothill zone to the alpine layer. All these characteristics, along with the narrowness of this canyon, result in the alteration of the vegetation zones because of temperature inversion. As for woodlands, more than one-third are beech-silver fir mixed woods while one-fourth corresponds to holm oak woods. However, what stands out is that it gives shelter to both the hazel-mixed woods and nearly all the holm oak woods in the Park.

1 Amelanchier ovalis Medik. (Rosaceae)

Snowy Mespilus. Amélanchier. Guillomo. *Senera, griñolera*
A. rotundifolia (Lam.) Dum.-Courset, **A. vulgaris** Moench

SECT.: B O T A E P C. ALT.: 650 - 2000 m. NP dec. Submed. C

In spring, when this submediterranean bush is in full blossom, the warmest areas in the Park such as the slopes of Andecastieto, at the entrance of Ordesa, are covered with a display of white. The name of the genus refers to the sweet honeyed taste of its ripe fruit, whilst that of the species has to do with the oval leaf shape.

Eco.: It is a characteristic plant of *Quercus* forests (*Q. subpyrenaica* and holm oak) and scrubs dominated by boxwood, bearberry or *Echinospartum horridum*, mainly on calcareous and loamy soils.

Loc.: Cardal ravine, Otal (B); Los Navarro bridge, Gradas de Soaso (O); Bordas ("barns") de Aso, Mondoto (A); Escuain pueblo, Revilla (E); Parador, Zapatierno (P).

2 Arctostaphylos uva-ursi (L.) Spreng. (Ericaceae)

Bearberry. Busserole, raisin-d'ours commun. Gayuba. *Buchareta, bucheta, manzanetas de pastor*

SECT.: B O T V A E P C. ALT.: 950 - 2300 m. Ch.
Plurireg. (Bor.-alp. y Med. Mount.) F

Procumbent evergreen shrub. Curiously, both the genus name, *Arctostaphylos*, and that of the species, *uva-ursi*, have the same meaning, that is bear-berry, the former coming from Greek while the latter is its latinized form.

Another plant in the same genus, the alpine bearberry, ***A. alpinus***, with leaves withering in autumn and black ripe fruit (small image), occurs in Bujaruelo valley.

Eco.: It carpets mountainsides, ridges and various types of open luminous woodlands on stony soil, both in montane and subalpine zones.

Loc.: Turbón, Lapazosa ravine (B); Andecastieto, Cotatuero (O); San Úrbez, la Miguasa Pass (A); La Loresa, toward los Mallos bridge (E); Espierba southern slope, La Larri (P).

3 Buxus sempervirens L. (Buxaceae)

Box, Boxwood. Buis. Boj. *Bucho, buxaco*

SECT.: B O T V A E P C. ALT.: 700 - 1900 m. MP everg. Submed. CCC

The box or *bucho* is an evergreen shrub, as its specific name states, although its leaves become reddish orange in winter as they accumulate pigments called anthocyanids that protect them from cold. Its hard wood is used to make some decorative craft utensils. It is also employed for hedges as it is highly tolerant of close shearing and produces newly emerging shoots easily. It has been used in popular medicine to reduce fever or as a hair loss treatment, also as a purgative, etc., although it is quite toxic, even fatal, as it contains some alkaloids. Some local place-names such as Buchaco or Bujaruelo, show that it grows in large quantity there.

Eco.: It is either the most abundant or the most ecologically adaptable shrub in the Pyrenees as it thrives in various types of forests such as holm oak and *Quercus subpyrenaica* woodland, mixed woods, pinewoods, beechwoods and silver fir woods ranging from the lowlands to the upper montane zone on stony calcareous soils. It makes up large extents of successional shrub communities (box scrub or *bujedos*), either on its own or mixed with *Echinospartum horridum* and other shrubs. It also forms the scrubby permanent vegetation in barren spots.

Loc.: Oncins bridge, Otal (B); Andecastieto, La Canal ravine, Turieto (O); Fanlo-Nerín southern slope, San Úrbez, La Pardina (A); La Valle, Castillo Mayor, Revilla (E); Tormosa ravine, La Larri, Parador (P).

4 Cardamine heptaphylla (Vill.) O.E. Schulz (Brassicaceae)

Toothwort. Cardamine à sept folioles, dentaire pennée. Dentaria

Dentaria pinnata Lam.

SECT.: B O A E P. ALT.: 750 - 1770 m. G. Eur. S E

Perennial herbaceous plant with palm-like (hand-shaped) leaves made up of 7 leaflets, as the name of the species indicates, with white or pinkish flowers that bloom in early spring, before the trees put forth new leaves.

Eco.: This is species characteristic of fresh shady spots with humus-rich soil, growing at the bottom of ravines or in wet woods, particularly in beechwoods and hazel thickets.

Loc.: Oncins bridge, Ordiso (B); Bosque de las Hayas, Senda de Cazadores (O); Las Cambras gorge, La Pardina ravine (A); Airés ravine (E); bottom of Pineta valley (P).

5 Cypripedium calceolus L. (Orchidaceae)

Lady's Slipper Orchid. Sabot-de-Vénus, Soulier-de-la-Vierge
Zapatito de la Virgen, zueco de Venus

SECT.: O P. ALT.: 1200 - 1375(1900) m. G. Circumb. (Eur.) RI

This plant bears the largest flowers among the Pyrenean orchids, with their unmistakable slipper-shaped labellum. It is pollinated by insects including bees, wasps and ants (Hymenoptera), that touch the stamens (it lacks the pollinia) with their bodies when trying to escape from the labellum. Its elliptical, slightly sheathing leaves bear glandular hairs that may irritate the skin. The rhizomes have been used in naturopathy, especially in America. The Latin name refers to Venus or Aphrodite's footwear, the goddess of Cyprus.

Eco.: It grows in shady mixed forests, beechwood clearings and subalpin shrubs with Pyrenean willow.

Loc.: Only three populations of this rare orchid have been found so far i the Aragonese Pyrenees: Sallent de Gállego, Ordesa and Pineta, all thes constituting the absolute SW distribution limits of the species in Europe

6 Daphne mezereum L. (Thymeleaceae)

Mezereon, February Daphne. Daphné bois-gentil, Daphné joli bois. Mezereón. *Matacrabas*

SECT.:.: B O A E P. ALT.: (975)1220 - 2180 m. NP dec. Eur. F

Its intensely fragrant pink flowers are produced on woody bar stems before the new leaves appear. The red fruit (small image can be used to dye the wool yellow. Their toxicity is well-known a they provoke serious digestive trouble and may even cause death just as the Aragonese name states ("matacabras" meaning goat-ki ller). The genus was named in honour of Dafne, daughter of river god Peneus, who was transformed into a laurel-tree, from whic another related species also present in the Park, the Spurge-laure (*Daphne laureola*) takes its name (small image).

Eco.: It grows in montane and subalpine we forests: beech, silver fir, mixed and moun tain pine woods.

Loc.: Santa Elena hermitage, Otal (B); Lañ Caballo, Gradas de Soaso (O); La Pardina ra vine, Fuen Blanca (A); La Valle, trail to Revi lla (E); Parador, Faja Tormosa (P).

7 Echinospartum horridum (Vahl) Rothm. (Fabaceae)

Genêt horrible, genêt très épineux. Erizón. *Brinzón, escarpín*

Genista horrida (Vahl) DC.

SECT.: B O T V A E. ALT.: (650)1000 - 1950(2100) m. NP(Ch).Pyr.-Ceven. RR

This shrub is greatly beneficial to our mountain landscapes for it prevents many deforested slopes from being eroded as it is an efficient colonizing plant. It also helps soil regeneration thanks to nitrogen fixation bacteria which live symbiotically in the roots, just as is the case with other leguminous species. Its scientific name means that this spiny plant stands your hair on end if you happen to touch it. One of its Aragonese names, «brinzón», refers to the locals of Nerín —a hamlet situated in the Park peripheral zone in Vió valley— who are called «brinzoneros».

Eco.: This spiny cushion-like shrub is in full blossom at the end of June when entire, deforested, repeatedly-set-on fire, sunny-facing mountainsides just turn yellow. It also occupies abandoned fields where *Quercus subpyrenaica* and pine woodland occurred long ago.

Loc.: Ordiso, Bujaruelo Pass (B); Soaso, Las Cutas southern slope (O); Sercué, La Minguasa Pass (A); O Furicón, Foratarruego, La Loresa (E).

8 Gentiana lutea L. subsp. montserratii (Vivant) O. Bolòs & Vigo (Gentianaceae)

Montserrat's Great Yellow Gentian. Gentiane jaune du P. Montserrat. Genciana de Pedro Montserrat

G. montserratii Vivant

SECT.: B O A E P. ALT.: 1300 - 1760 m. H. Pyr. R

A plant with many ethnobotanical usages as the root, after being boiled, is used to elaborate aperitifs such as the well-known Chartreuse. In Sobrarbe it was employed to stimulate livestock appetite, mixed with quinine and wormwood (*Artemisia absinthium*). It has been used to lower blood pressure (mermasangre meaning "blood-reducer" in Aragonese) although this therapeutic effect has not been medically verified. The Gentian violet, a dye, is harvested from its rhizome. This plant is dedicated to the distinguished botanist from Jaca, Pedro Montserrat Recoder, born in Mataró (Barcelona).

Eco.: This subspecies prefers calcareous soils at the fringe or the clearings of forests, also various sorts of pastures and it even grows at the foot of stony outcrops.

Loc.: Cotatuero, Salarons, Gradas de Soaso (O); Castillo Mayor (E); Estiba de Espierba, bottom of the Pineta valley (P).

9 Goodyera repens (L.) R. Br. (Orchidaceae)

Creeping Lady's Tresses, Dwarf Rattlesnake Plantain. Goodyère rampante. Goodyera

SECT.: O A E P C. ALT.: 1100 - 1600(1900) m. G. Eur. **E**

This orchid's tiny flowers of look closed, therefore insects have some trouble reaching their nectar. When trying to get into them, they end up touching the rostellum and so removing the pollinia, which stick to their body; the rostellum rises, setting apart from the labellum and clearing a way towards the nectar and the stigma zone. The plant is named after the 17th-century English botanist John Goodyer whereas the species name refers to its creeping stoloniferous rhizome. It has been assigned various healing properties, even though they have not been scientifically confirmed.

Eco.: Delicate forest orchid that takes root in the humus in silver fir or mossy pine woods.

Loc.: Turieto, Cotatuero (O); Yaga gorge, Angonés ravine (E); trail to Montinier and Añisclo mountain Passes (P).

10 Helleborus foetidus L. (Ranunculaceae)

Stinking Hellebore, Dungwort, Bear's Foot. Ellébore fétide. Eléboro fétido. *Chigüerre*

SECT.: B O T A E P C. ALT.: 700 - 2160 m. Ch. Eur. **CC**

Perennial plant of dull, dark-green, palm-like (digitate) leaves; it stinks when touched as its name says. Its yellowish-green flowers open out at the height of winter. It is a very toxic plant that has been used to treat heart trouble. In Sobrarbe it was used in a topical way to heal the swollen neck ganglia.

Eco.: It is found in forest clearings, scrub, stony pastures, screes, etc., up to the subalpine zone all over this protected area where it reaches its altitudinal limit in the Aragonese Pyrenees.

Loc.: Cardal ravine, Salto del Pich (B); Turieto, Cola de Caballo (O); Bordas ("barns") de Aso, Cuello Arenas (A); Los Mallos bridge, La Valle (E); Montaspro, Tormosa ravine (P).

11 Helleborus viridis L.

subsp. **occidentalis** (Reut.) Schiffn. (Ranunculaceae)
Green Hellebore. Ellébore vert. Eléboro verde. *Matapullo*
H. occidentalis Reut.
SECT.: B O T A E P. ALT.: 1025 - 1935 m. G. Lateatl. **F**

This hellebore is very similar to the one mentioned before, but its leaves are shining green, also with broader leaflets. It contains some toxic active constituents as the stinking hellebore and as a matter of fact they were used in the same way.

Ɛco.: This species has narrower ecological requirements than the previous one so it just occurs in wet deciduous forests.

.oc.: Ordiso bridge, Oncins bridge (B); Laña Caballo, Faja Pelay (O); Plana Canal, Capradiza ravine (A); Yaga canyon, La Valle (E); Tormosa ravine, El Felqueral (P).

12 Hepatica nobilis Schreb. (Ranunculaceae)

Liverwort, Kidneywort, Pennywort. Anémone hépatique, hépatique à trois lobes. Hepática, trébol dorado. *Hoja de tres picos*
Anemone hepatica L., **H. triloba** Chaix
SECT.: B O T V A E P C. ALT.: 800 - 2010 m. H. Eur. **CC**

t is one of the earliest-flowering species, with violet-blue or white petals. Its name refers to the leaf-shape which looks like a three-lobed liver. According to the "theory of the signals", plants would show their possible usages by means of their shape so this one was used for curing liver ailments. In Bielsa and elsewhere in the Pyrenees its leaves were used to make a poultice which was intended to cure injuries while in Fanlo this was made by frying the freshly picked plant and then rubbing the injured parts.

Ɛco.: It thrives in all kind of forests, including holm oak, pine and beech woodland all over the Park, even in the subalpine zone.

.oc.: Faixa O Cordiador, San Nicolás de Bujaruelo (B); Laña Caballo, Bordas ("barns") de Sanguino (O); San Úrbez, Betosa ravine (A); La Valle, Surgencia del Yaga (E); Parador; Tormosa ravine (P).

13 Moneses uniflora (L.) A. Gray (Pyrolaceae)

One-flowered Wintergreen, Single Delight. Pirole à une fleur
Peralito unifloro

Pyrola uniflora L.
SECT.: B O T A E P C. ALT.: 1050 - 1800(2000) m. H(G). Bor.-alp. E

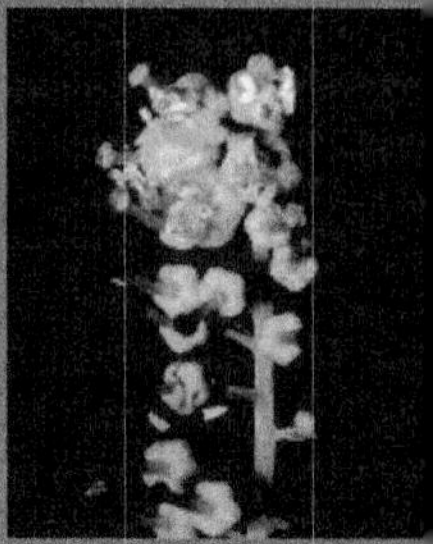

Both the genus and the species name refer to the fact that it produces a single white flower per year, appreciably large compared to the whole plant itself. The Spanish vernacular name refers to the fruit shape, which looks like a tiny pear.

Orthilia secunda (small images), the Sidebells Wintergreen, whose stems produce several flowers all facing the same direction, occurs in this same environment.

Eco.: This small plant takes root in the humus of wet forests in the shade of montane and subalpine pinewoods or silver fir woods.

Loc.: Turbón (B); Turieto, La Canal ravine (O); Añisclo canyon (A); Los Mallos bridge, Castillo Mayor (E); Ronatiza, Pineta hermitage (P).

14 Neottia nidus-avis (L.) L.C.M. Richard (Orchidaceae)

Bird's Nest Orchid. Néottie nid-d'oiseau. Nido de pájaro

SECT.: B O A E P. ALT.: (1050)1235 - 1660 m. G. Eur. E

This orchid lacks chlorophyll the entire plant is dirty whit or brownish and has thick en tangled roots that are remi niscent of a bird's nest, whicl is the reason for both its Lati and vernacular names.

Eco.: It grows in wet humus-ricl forests such as beech, silver fir c mossy pine woods.

Loc.: Murallas de la Gatera (B) Senda de los Cazadores, Bosqu de las Hayas (O); La Pardina ravin (A); Los Mallos bridge; Faja Colga da (E); trail to Añisclo mountai Pass, Cornato (P).

15 Oxalis acetosella L. (Oxalidaceae)

(Common) Wood Sorrel. Oseille des bois, oxalide petite. Acederilla. *Aleluya, vinagreta*

SECT.: B O T A E P. ALT.: (750)950 - 2000 m. H. Lateeur. C

It is an early-flowering plant with clover-shaped leaves which crease when they do not have enough humidity. The English and some of the Spanish vernacular names and also the Latin name of the species all refer to the sour taste of the leaves. This plant turns out to be toxic in large doses. It was used as a remedy against scurvy and as a diuretic as well.

Eco.: It carpets wet shady woodlands such as beech, silver fir and pine woods on non-calcareous soils.

Loc.: Ordiso, Santa Elena bridge (B); Turieto, Faja Pelay (O); San Úrbez, La Mirona ravine (A); Surgencia del Yaga, Sestrales (E); Es Camons, Tormosa ravine (P).

16 Paris quadrifolia L. (Liliaceae)

Herb Paris, True Lover's Knot. Parisette à quatre feuilles. Ojos de zorra. *Uba de raposa*

SECT.: B O P. ALT.: 1200 - 2000 m. G. Eur. E

This forb has got four leaves, all of them horizontally arranged and inserted at the same point of the stem, as its Latin name says. The Aragonese vernacular name meaning "fox grape" comes from the fruit, a black berry. This toxic plant slows down the heartbeat if eaten.

Eco.: It grows on soils rich in decomposing organic matter within beech and mixed woods.

Loc.: Ordiso, Santa Elena bridge (B); Faja Pelay, Laña Caballo (O); Es Camons, Tormosa ravine (P).

17 **Primula veris** L. (Primulaceae)

(Common) Cowslip. Primevère officinale. Primaveras. *Cacaleta*

SECT.: B O T V A E P. ALT.: 700 - 2100(2250) m. H. Eur. **C**

This is an early-flowering species, as its scientific name points out, which produces a group of intensely yellow flowers at the top of a short stem. Its freshly-picked leaves are edible after being boiled. In Sobrarbe it is used to make a poultice to treat bruises, inflammation, etc., and in fact, some natural pain-killers, among other substances, have been isolated.

Eco.: It inhabits montane forests, downy oak, pine and beech woods above all. It also occurs in pastures, meadows, fresh hedges and box scrub everywhere in the Park.

Loc.:Gabieto ravine (B); Los Navarros bridge (T); Cotatuero, Gradas de Soaso (O); road to Añisclo (A); Escuaín (E); Cirque of Pineta, La Larri (P).

18 **Pulsatilla alpina** (L.) Delarbre subsp. **font-queri** M. Laínz & P. Monts. (Ranunculaceae)

Font-Quer's Alpine Anemone. Anémone blanche de Font-Quer. Flor del viento, *Anémone*

SECT.: B O T V A E P C. ALT.: 1100 - 2400 m. H. NE Iber.-Pen. **CC**

This anemone produces single white flowers suffused with blue. Its fruit are topped with a feathery appendix which help them fly and disperse with the wind, as the genus name points out. It has got hairy dissected leaves. This plant, which has been used as an antibiotic, is dedicated to the eminent botanist from Lleida Pío Font i Quer. The field layer in some of the beechwoods within the Park is carpeted by a related species, the Wood Anemone (***Anemone nemorosa*** L., small upper image) and Yellow Wood-Anemone (***A. ranunculoides*** L., small below image).

Eco.: North-facing screes, stony grasslands and open subalpine pinewoods.

Loc.: Duáscaro ravine, Plana Cuasta (B); Faja Pelay, Soaso (O); Sierra Custodia, Cuello Arenas (A); Los Mallos bridge, La Consusa ravine (E); Faja Tormosa, Cirque of Pineta (P).

19 Pyrola chlorantha Sw. (Pyrolaceae)

Great Flowered Wintergreen. Pirole à fleurs verdâtres, pirole verdâtre. Peralito

P. virens Koerte

SECT.: O T A E P. ALT.: 1000 - 1680 m. H. Plurireg. (Bor.-alp. y Eur.) F

This small herbaceous plant has got rounded leaves at the base of the stem. The name of the genus refers to the leaves, which are reminiscent of those on a pear tree (*Pyrus communis*). The epithet reveals the yellowish green colour of the flowers.

Eco.: This is a partly saprophyte, as it feeds on humus, and inhabits montane coniferous wet forests such as Scots pine and silver fir woods.

Loc.: Turieto, Cotatuero (O); Faja Colgada, Yaga gorge (E); Tormosa ravine, Cornato (P).

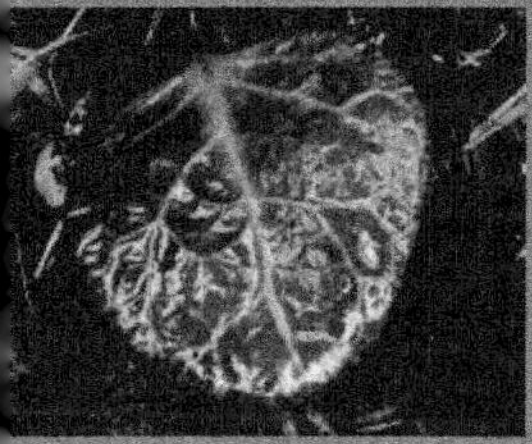

20 Rhododendron ferrugineum L. (Ericaceae)

Alpenrose, Rusty-leaved alpenrose. Rhododendron ferrugineux. Azalea de montaña. *Farnucera, barrabón*

SECT.: B O T V A E P C. ALT.: 950 - 2550(3040) m. NP everg. Alp . F

High-mountain shrub with rose-like flowers as stated by the genus name, which comes from the Greek word *rhodo*, and *dendron*, tree. Its epithet refers to the reddish ferruginous hairs that cover the back of the leaves, which emit a strong smell of basil when rubbed. A fungus called *Exobasidium rhododendri* (small image) brings about those reddish or yellowish galls which often appear on its leaves.

Eco.: It forms subalpine heaths on siliceous soils. It also characterizes the understorey of mountain pine forests, or rarely subalpine beechwoods (as in Faja Pelay). Curiously enough, this is a plant sensitive to cold and needs snow insulation.

Loc.: We have found this shrub in several abyssal locations within the Park with temperature inversion, growing on siliceous sandstone outcrops. We have also seen it at La Munia above 3,000 metres, its altitudinal limit in the Pyrenees. Vilá ravine, Cardal ravine (B); Faja Pelay, Turieto Alto (O); Sestrales, El Portiello (A); Los Mallos bridge, Airés ravine (E); Faja Tormosa and La Larri (P).

21 Rosa pendulina L. (Rosaceae)

Alpine rose. Églantier à fruits pendants. Rosal alpino
R. alpina L.

SECT.: B O V A E P. ALT.: (1300)1500 - 2340 m. P dec. Alp. F

This species has got a hanging appearance and bears just a few prickles unlike most roses. Its name refers to the pedicels that hold the flowers, which recurve when the fruit develops. Like in all roses, the fruit is very astringent because of its high tannin content and has also been used as a source of vitamin C.

Eco.: It embellishes the silver fir woods and subalpine mountain pine woods with Alpenrose, together with the corresponding successiona shrubs, and sometimes can be found in megaforbs as well.

Loc.: Faja Pelay, Turieto (O); La Pardina ravine, Sestrales (A); Faja Colgada, Castillo Mayor (E); Faja Formosa, Estiba de Espierba (P).

22 Rosmarinus officinalis L. (Lamiaceae)

Rosemary. Romarin. Romero. *Romé*

SECT.: A. ALT.: 750 - 1030 (1300) m. NPeverg. Med. F

The rosemary is an branched erect shrub which blossoms in mid winter. It is well-known because of its medicinal properties, as th Latin epithet *officinalis* indicates, meaning it is used in pharmac and also as a seasoning and garden plant.

Eco.: This is a plant sensitive to cold that forms successional scrub o calcareous soils, replacing former holm oak woods.

Loc.: There is only a small amount in the Park in Añisclo, wher it enters through the Cinca va lley, which constitutes one of it northernmost locations in th Pyrenees.

23 Salix daphnoides Vill. (Salicaceae)

Violet Willow. Saule faux daphné, saule à bois glauque. Sauce

SECT.: P. ALT.: 1170 - 1300 m. MP dec. Bor.-alp. RR

Shrub or small tree characterized by its branches being covered with a sort of wax which lend them a glaucous colour. The species name refers to the leaf-shape resemblance to that of the spurge-laurel (*Daphne laureola*) and the bay-tree or laurel (*Laurus nobilis*), as well.

Eco.: It is a part of the willow copses occurring on riversides and mixed woods, where it sometimes forms monospecific groups with some specimens that may reach 10 metres in height.

Loc.: One of the few populations of this willow in the Pyrenees is found in Pineta, where it still persists as a glacier relic.

24 Scilla lilio-hyacinthus L. (Liliaceae)

Pyrenean Squill. Jacinthe des Pyrénées. Jacinto de bosque

Tractema lilio-hyacinthus (L.) Spet

SECT.: B O A E P. ALT.: 995 - 1870 m. G. Atl. E

This bulbous perennial blooms early in the season before beech trees put forth new leaves. It produces beautiful star-like blue flowers that form loose racemes. The name of the species comes from its broad soft leaves that look like those of hyacinths, and which decompose quickly after the plant has fruited, so it is hard to be noticed in midsummer.

Eco.: It inhabits beechwoods with rich calcareous soil, sometimes forming large clusters.

Loc.: Ordiso, Oncins bridge (B); Bosque de las Hayas, Senda de Cazadores (O); La Pardina ravine, San Úrbez (A); Los Mallos bridge, Surgencia del Yaga (E); Tormosa ravine, bottom of the Pineta valley (P).

FORESTS CLEARINGS and MEGAPHORBS

When a clearing is torn within a forest because of an avalanche or an occasional tree fall, forest disruption and the ensuing light gap activate a natural regeneration process launched by sun-loving, colonizing, opportunistic species.

On the other hand, at the foot of humid cliffs or in forest clearings cut by ravines, usually rich in organic matter, tall large-leaved herbaceous species, known as megaphorbs, proliferate.

25 **Aconitum anthora** L. (Ranunculaceae)

Anthora, Yellow Monkshood, Healing Woolfsbane. Aconi anthore. Acónito amarillo. *Tora*

SECT.: B O A E P. ALT.: (1300)1500 - 2020 m. H. Alp. F

It is a rhyzomatous herbaceous plant which bears racemes of yellov flowers shaped like a helmet. It is very toxic and causes poisonin to both animals and people.

Eco.: It can be found at the foot of fresh limestone cliffs and in mois pastures as well.

Loc.: Picamartillo, Salto del Pich (B); Faja Racón (O); La Pardina ravine Capradiza ravine (A); Castillo Mayor (E); Tormosa ravine, El Felqueral (P)

26 **Aconitum napellus** L. subsp. **vulgare** Rouy & Foucau (Ranunculaceae)

Aconite, Monkshood, Wolfsbane. Aconit commun. Acónito verdegambre. *Toara*

SECT.: O P. ALT.: 1225 - 2120 m. H. Alp. E

This tall stout plant up to 2 metre high has got blue helmet-shape flowers gathered in racemes. Th Latin surname shows that its root actually its rhizome, is shaped lik a turnip. It is also a very toxic plan which causes death by cardio-respi ratory collapse and which was use to commit murder.

Eco.: It grows in wet pastures together with other tall herbs (megaphor bs), where livestock or cattle stay for a rest and therefore organic matte accumulates, that is to say nitrogen-rich spots.

Loc.: Near La Pradera (O); Faja Tormosa, las Inglatas, Estiba de Espierb (P).

27 Atropa belladonna L. (Solanaceae)

Bella Donna, Deadly Nightshade. Belladone. Belladona. *Pech*

SECT.: B O T E P. ALT.: 1100 - 1750 m. H. Eur. E

This is a very toxic plant which affects the heart rate and accordingly one of its active ingredients, atropine, is an essential ingredient in some medicines. *Atropa* means "cruel or inexorable" in Greek, and refers to the deadly properties of this plant. Roman women used it to dilate their pupils as they believed that by doing so they would become more desirable at men's eyes.

Eco.: It shows up in nitrified clearings opened by avalanches or falling trees within moist forests.

Loc.: Santa Elena bridge, Oncins bridge (B); Turieto Alto, Cotatuero (O); Yaga gorge, Tabacoy (E); Parador, Estiba de Espierba (P).

28 Chenopodium bonus-henricus L.

(Chenopodiaceae)

Lincolnshire Spinach. Chénopode Bon-Henri. Zurrón. *Sarrión*

Blitum bonus-henricus (L.) C.A. Mey.

SECT.: B O V A E P. ALT.: (1300)1500 - 2550 m. H. Eur. F

A very well-known plant to the locals of these mountains as it is consumed by both livestock and people. The leaves are also employed to make a poultice to cure injuries. The Latin name refers to the shape of a goose foot (*chenos* in Greek) of the leaves in some of the species of this genus. This particular species is dedicated to King Henry IV.

Eco.: The Perennial Goosefoot abounds in sheepfolds with nitrogen-rich soil, where it is sometimes the main species.

Loc.: Cardal ravine, Plana Cuasta Pass (B); Góriz refuge, Gradas de Soaso (O); Cuello Arenas, Cave of Pájaro Muerto (A); Cueva Foratata, Chasm of La Bufona (E), Faja Tormosa, El Felqueral (P).

29 **Digitalis lutea** L. subsp. **lutea** (Scrophulariaceae)

Yellow Foxglove. Digitale à petites fleurs, digitale jaune. Dedalera amarilla. *Tuera amarilla*

SECT.: B O A E P. ALT.: 700 - 1860 m. H. Latesubmed. **F**

A very toxic plant which contains, among other active ingredients, digitaline, used to treat heart failure and other coronary ailments. The Spanish name refers to the resemblance of its yellow flowers to a thimble ("dedal").

The Flowglove, ***D. purpurea*** (see image below) appears on siliceous substratum in the Park's peripheral area.

Eco.: It is found in fresh forest clearings and screes with soil rich in organic matter.

Loc.: Cotatuero, Arripas waterfall (O); San Úrbez, Fuen Blanca (A); Angonés ravine, Los Mallos bridge (E); El Felqueral, Tormosa ravine (P).

30 **Leuzea centauroides** (L.) Holub (Asteraceae)

Rhapontique à feuilles de centaurée

Rhaponticum centauroides (L.) O. Bolòs non G. Don, **Rh. cynaroides** Less., **Stemmacantha centauroides** (L.) Dittrich

SECT.: B O A E P. ALT.: (1200)1350 - 1900(2110) m. H. Pyr. C **E**

In the heart of the summer, this tall, stout herbaceous plant produces large bluish heads that are similar to artichokes, whilst the epithet centauroides refers to its notable resemblance to some species of *Centaurea*.

Eco.: This is a plant which requires rather deep, rich soils, regardless if they are stony, either in pastures, meadows or megaphorbs and chiefly on limestone.

Loc.: Ordiso (B); Carriata, Soaso (O); La Pardina ravine, Fuen Blanca (A); Castillo Mayor (E); track to La Larri (P).

31 Lilium martagon L. (Liliaceae)

Turk's Cap Lily, Martagon. Lis martagon. Martagón. *Azucena de puerto*

SECT.: B O V A E P. ALT.: (650)950 - 2000 m. G. Eur. F

One of the most beautiful flowers in our mountains, pink or purple-coloured, it was used for curative or decorative purposes in a number of valleys in the Pyrenees.

Eco.: It occurs within or on the border of moist forests such as pine, beech, silver fir or mixed woods.

Loc.: Oncins bridge, Salto del Pich (B); Turieto, Senda de Cazadores (O); Capradiza ravine, San Úrbez (A); Los Mallos bridge, Faixa Billa (E); Tormosa ravine, Parador (P).

32 Lilium pyrenaicum Gouan (Liliaceae)

Pyrenean Lily, Yellow Turk's Cap Lily. Lis des Pyrénées. Azucena del Pirineo

SECT.: B O A E P. ALT.: 1000 - 1700 m. G. Late Pyr.-Cant. E

This lily is scarcer than martagon, and differs in that it bears spotted yellow flowers whilst the leaves are scattered along the stem, instead of being in whorls as it occurs with its relative. It is endemic to both the Pyrenees and the Cantabrian Range.

Eco.: It appears in moist pastures and forest fringes.

Loc.: Burguil bridge, Oncins bridge (B); Cotatuero, Bordas ("barns") de Salarons (O); Capradiza ravine (A); trail to Surgencia del Yaga, Angonés ravine (E); Parador, El Felqueral (P).

33 Rubus idaeus L. (Rosaceae)

Raspberry. Framboisier. Frambueso. *Chordonera*

SECT.: B O T A E P. ALT.: (950)1175 - 2180 m.
P dec. Plurireg. (Eur. y Bor.-alp). F

A shrub with leaves that are downy-whi te underneath and whose prickles ar weaker than those of other brambles Raspberries can be both eaten for des sert and used to prepare jam. Its leave have well-tested astringent and diureti properties whilst the fruit is laxative an refreshing. The Latin epithet refers t the isle of Ida, in The Troad (Anatolia).

Eco.: It thrives in clearings and opening within wet forests due to avalanches or fa lling trees, usually if the soil is rich in orga nic matter, together with the goat willow.

Loc.: Oncins bridge (B); Salarons, Faja Pela (O); Bordas ("barns") de Aso, San Úrbez (A) Los Mallos bridge, Yaga gorge (E); Tormos ravine, trail to Añisclo Pass (P).

34 Sambucus racemosa L. (Caprifoliaceae)

Red Elderberry. Sureau à grappes. Saúco rojo. *Sabuquer royo*

SECT.: B O A E P. ALT.: (845)1100 - 1950 m. MP dec. Lateeur. R

Tree-like shrub with sof wood whose inflorescenc resembles a raceme, as it surname specifies. It als differs from other member of the same genus in its re fruit.

Eco.: Like raspberries, it lives in forest clearings torn by avalanches o falling trees.

Loc.: Gabieto ravine, Los Navarros gorge (B); Cotatuero, Turieto (O); L Pardina ravine, Aforo (A); Castillo Mayor, La Garganta ravine (E); Tormos cave, El Felqueral (P).

35 Saponaria ocymoides L. (Caryophyllaceae)

Rock Soapwort, Tumbling Ted. Saponaire des rochers. Jabonera rastrera

SECT.: O T A E P C. ALT.: 750 - 1470(1880) m. Ch(H). Lateeur. R

This is a creeping plant that bears pink flowers on much-branched floriferous stems. Its Latin name comes from some of the species of the same genus having been used as soap, particularly the common soapwort *Saponaria officinalis* (small image below) which has been found in Revilla.

Eco.: It inhabits dry sunny spots such as the successional scrub that derives either from holm oak or Quercus subpyrenaica woods.

Loc.: Los Navarros bridge, La Canal ravine (O); San Úrbez, trail to Gallisué (A); Escuaín, trail to Surgencia del Yaga (E); Espierba (P).

36 Valeriana pyrenaica Ramond (Valerianaceae)

Pyrenean Valerian. Valériane des Pyrénées. Valeriana del Pirineo. *Yerba de los aguachals*

SECT.: B O A. ALT.: (950)1300 - 1840 m. H. Late-Pyr.-Cant. R

This is a fairly tall, large-leaved herbaceous plant, what botanists call a megaphorb, and it bears pink fragrant flowers. Despite its Latin epithet, it can also be found beyond the Pyrenees, in the Cantabrian Mountains and the Iberian System.

Eco.: We can find this species growing in humid sites by streamsides or brooks, as its Aragonese name indicates.

Loc.: Burguil bridge, Ordiso (B); La Vaqueriza (O); La Ripareta, San Úrbez (A).

PASTURES

A long-time interaction between both wild and domestic herbivores and vegetation has given rise to the evolution of a series of biological types and plant communities specifically adapted to grazing. Four types of livestock —cows, sheep, horses and goats—, together with the Pyrenean ibex and chamois, have been grazing traditionally in our mountains, each one showing different preferences to pasture. All these, together with geological, microclimatic and topographic variations (elevation, slope exposure, inclination) cause the pastures to be the most diverse plant communities in our landscape.

Thus, we have Mediterranean pastures in lower Añisclo valley, rich in woody, aromatic species, and looking already green at the end of the winter but quickly withering in midsummer; besides fresh montane pastures used by stock in spring and finally high-mountain pastures or "estivas" that are grazed only during the short Pyrenean summertime. The latter show large dissimilarities in their floristic compositions depending on the substratum (calcareous or siliceous), soil depth, stony nature, topographic position (southern or northern slopes, hollows), snow or ice melting patterns, etc.

37 Achillea millefolium L. (Asteraceae)

Common Yarrow. Achillée millefeuille. Milenrama. *Merma sangre*

SECT.: B O T V A E P C. ALT.: 940 - 2230(2350) m. H. Eur. C

This herbaceous plan takes its Latin surnam from its much-divided almost feathery leaves which are also very aro matic. The name of th genus refers to the Gree hero Achilles, that succe eded in healing Telephus wounds by using the ya rrow. In fact it has bee used to cure both huma and livestock ailments and also for folk remedies.

Eco.: It thrives in fresh meadows and mesophilous pastures of the monta ne and subalpine zone.

Loc.: Turbón, El Cebollar ravine (B); Casa Bergés, Gradas de Soaso (O) Cuello Arenas, Las Traviesas spring (A); Foratata cave, Tozal de San Vicer da (E); Faja Tormosa, track to La Larri (P).

38 Antennaria dioica (L.) Gaertn. (Asteraceae)

Mountain Everlasting, Castfoot, Cudweed. Antennaire dioi que. Pie de gato. *Cola de fuina*

SECT.: B O T V A E P C. ALT.: (1280)1500 - 2450(2710) m. Ch. Latebor.-alp. C

A lay person may mistake this species for the edelweiss (*Leontopodium alpinum*), as its flower heads are also white but much smaller. The name of the genus, *Antennaria*, alludes to the hairs of the male capitula which are thickened above like the insect antennae. Its surname states that some specimens are male while others in contrast are female. It has been used to heal some respiratory ailments and as an anti-inflammatory remedy.

Eco.: It grows in acidophilous pastures, mainly in the subalpine and alpine belts.

Loc.: Lapazosa ravine, Espelunz (B); Faja Pelay, Monte Perdido (O); Sierra Custodia, Plana Canal (A); Foratarruego, Montaña de Sesa (E); trail to Añisclo Pass, Estiba de Espierba (P).

39 Aphyllanthes monspeliensis L. (Liliaceae)

Oeillet bleu de Montpellier. Junquillo. *Chunqueta*

SECT.: B O A E P. ALT.: 690 - 1470 m. H. Med. W (Submed). **E**

Its Spanish vernacular names are very accurate indeed as this plant resembles a small rush ("junco"). Its leaves are reduced to sheaths that wrap the hollow, slender stems topped with pretty blue flowers bearing six petals and six stamens. The Latin epithet, *monspeliensis*, refers to the city of Montpellier, from where the species was first described. It is eaten by livestock in spring, when it is still tender. It has been used in popular medicine to improve circulation.

Eco.: A very characteristic plant of dry, sunny, sub-Mediterranean pastures on limestone (called "junquillares"), within rosemary or box scrub.

Loc.: Santa Elena gorge (B); La Ereta bridge, La Canal ravine (O); Bordas ("barns") de Aso, Gallisué (A); La Loresa, La Consusa ravine (E).

40 Asphodelus albus Mill. subsp. delphinensis (Gren. & Godr.) Z. Díaz & Valdés (Liliaceae)

White Asphodel. Asphodèle du Dauphiné. Gamón. *Abozos, albezones*

A. delphinensis Gren. & Godr., **A. pyrenaicus** Jord

SECT.: B O A E P. ALT.: 1080 - 2000(2200) m. G. Alp. **F**

This plant bears a single stem topped with a raceme of white flowers bearing showy stamens that jut out from them. The *abozos* (its Aragonese name) were harvested formerly in Soaso and Bujaruelo to feed swine. It has also been used in popular medicine and cosmetics, but its swollen roots are toxic.

Eco.: It abounds in heavily grazed hay meadows and pastures growing on nitrogen-rich, deep soils, as well as in box scrub, on rocky hillsides repeatedly set on fire, etc.

Loc.: Ordiso, Salto del Pich (B); Salarons, Gradas de Soaso (O); Bordas ("barns") de Aso, Capradiza ravine (A); trail to Surgencia del Yaga (E); Faixa Castiecho, Montaspro (P).

41 **Botrychium lunaria** (L.) Sw. (Ophioglossaceae)

Common Moonwort. Botryche lunaire. Botriquio

SECT.: B O V A E P C. ALT.: (1300)1600 - 3015 m. G. Bor.-alp. C

The presence of two different kind of fronds (equivalent of leaves in higher plants), one of them sterile and the other one bearing the spores, reveals the ancient origin of this curious little fern. In fact, its scientific name comes from the resemblance of the fertile blade to the branched stalk of a bunch of grapes.

Eco.: It colonises rather lime-devoid soils in a great variety of habitats, including dense, stony or wet pastures, snowbed communities and even grazed woodland, ranging from the montane to the subnival zone.

Loc.: The uppermost populations of this fern in the entire Aragonese Pyrenees have been found within the Park. Cardal ravine, Ordiso (B); Gradas de Soaso, Descargador (O); Cuello Arenas, Sestrales (A); Foratarruego, La Valle (E); Balcón de Pineta, Plana Fonda (P).

42 **Carex sempervirens** Vill. subsp. **sempervirens** (Cyperaceae)

Laiche sempervirente. Cárice siempre verde

SECT.: B O A E P C. ALT.: 950 - 2600(2950) m. H. Alp. CC

A tussock plant with long shiny-green leaves, as its Latin name implies.

Eco.: It grows either in stony and garland pastures or on rock ledges, mostly on north-facing slopes or snowbed communities. The subspecies *pseudotristis* has been found on siliceous soils in Bujaruelo valley, in the upper Ara.

Loc.: Las Comas ravine, Espelunz (B); Tobacor, Soaso (O); Cuello Arenas, Betosa ravine (A); Cuello Viceto, Foratarruego (E); Balcón de Pineta, Faja Tormosa (P).

43 Carlina acanthifolia All. subsp. cynara (Pourr. ex Duby) Rouy (Asteraceae)

Carline Thistle. Carline artichaut. Carlina. *Yerba de bruxas*
C. cynara Pourr.
SECT.: B O T E P. ALT.: (1000)1330 - 2220 m. H. Submed. W E

This stemless plant produces a large single cluster of flowers (capitulum) up to 15 cm in diameter, composed of tightly-packed yellow flowers ringed with a crown of golden leaves (the outer bracts). Its spiny leaves form a rosette and resemble those of the acanthus (*Acanthus mollis*), hence its Latin name. Although it has a number of applications in traditional medicine, there has been a widespread use of this plant in witchcraft in the Pyrenees: as a solar symbol it is hang on doors and windows of houses or stables, as a deterrent against lightning and to ward off witches and evil spirits as well.

Eco.: This *cardigaza* (as it is also referred to in Aragón) can be found in pastures above the tree line together with some forest clearings.

Loc.: Lapazosa ravine, Planabé (B); Carriata, Faja Canarellos (O); Angonés, La Loresa (E); trail to Añisclo Pass, Estiba de Espierba (P).

44 Carlina acaulis L. (Asteraceae)

Silver Thistle, Dwarf Carline Thistle. Carline à tige courte. Carlina angélica. *Cardo de puerto o de bruxa*

SECT.: B O T V A E P. ALT.: (950)1250 - 2470 m. H. Eur. CC

Unlike its relative, this thistle has silvery bracts, a smaller inflorescence with white florets suffused with purple and leaves more divided. In spite of its name indicating a lack of stem, some specimens may have one. Apparently *Carlina* may refer to Emperor Charlemagne, who used it in his struggle against epidemics. It was also used, as the previous species, to keep witches away from the Pyrenean houses.

Eco.: The *cardo de puerto* is one of the most attractive plants on the summer pastures frequented by livestock.

Loc.: Espelunz, Cardal ravine (B); Faja Pelay, Góriz (O); Cuello Arenas, Sierra Custodia (A); Castillo Mayor, Foratarruego (E); Faja Tormosa, Cirque of Pineta (P).

45 Centaurea scabiosa L. subsp. scabiosa (Asteraceae)

Greater Knapweed. Centaurée scabieuse. Cabezuela. *Yerb ojera*

C. cephalariifolia Willk.

SECT.: B O T V A E P. ALT.: (700)1150 - 1900(2200) m. H. Lateeur.

This knapweed is a perennia with divided leaves. It bear several heads ringed wit greenish bracts, ciliated o the margin, and bluish-purpl flowers. The outer flower are very attractive and large than the inner ones forming sort of star crown. Its usag in popular medicine to trea eye ailments is well known as its Aragonese name suggests ("ojera" meaning *of the eye*), or aga inst scabies, hence its scientific name.

Eco.: It can be found in sunny pastures, hay meadows, ditches, wast land, forest clearings, etc.

Loc.: Gabieto ravine, Lapazosa ravine (B); Andecastieto, Bordas ("barns" de Salarons (O); La Pardina ravine, Gallisué (A); La Valle, Escuaín (E); L Larri, bottom of the Pineta valley (P).

46 Crocus nudiflorus Sm. (Iridaceae)

Autumn Crocus. Crocus à fleurs nues. Azafrán silvestre

SECT.: B O A E P. ALT.: 1220 - 2380 m. G. Atl.

The bulb of this species is earnestly sought out by wild boars. It produces a bell-shaped lilac flower at ground level that proclaims the end of the summer, whilst the leaves appear afterwards, when the fruit is formed.

Within the Park we can find another plant of the same genus and also growing in pastures, the White Crocus, ***Crocus vernus*** subsp. ***albiflorus*** (small image), with white flowers blooming in spring, hence its name. It has been used in popular medicine as an antispasmodic and a sedative, as well as an abortive.

Eco.: It grows in various types of pastures above the timberline.

Loc.: Ordiso bridge, Cardal ravine (B); Punta Acuta overlook (O); Cuell Arenas, La Estiva (A); Escuaín, La Monesma (E); El Felqueral, Faja Tor mosa (P).

47 Doronicum grandiflorum Lam. subsp. grandiflorum (Asteraceae)

False Arnica. Doronic à grandes fleurs. Falsa árnica

Aronicum scorpioides DC., D. scorpioides Lam.

SECT.: B O A E P C. ALT.: (1320)1570 - 3065 m. H. Alp. F

This forb produces yellow flowers gathered in capitula, with ligules of the same colour. As its vernacular name implies, it has been mistaken many times for the true Mountain Arnica, *Arnica montana*, absent from the Park, which has entire opposite leaves, unlike *Doronicum* where they are dentate and alternate. As with all of the arnicas (either the true or the false ones) it has been used to heal wounds and sores and relieve bruises and rheumatism.

Eco.: It inhabits screes and the base of fresh shady cliffs, chiefly on limestone, and sometimes on high-mountain brook sides.

Loc.: In Ordesa valley it has been found at 1320 m, its lowermost locality in the Pyrenees known so far, while at Astazu Peak we have seen it living at 3065 m, the uppermost point in the entire mountain range. Ordiso, Los Batanes ravine (B); Mondarruego, Góriz (O); Punta Custodia, Añisclo Pass (A); La Valle, Montaña de Sesa (E); Faja Tormosa, Lake Marboré (P).

48 Dryas octopetala L. (Rosaceae)

Mountain Avens. Chênette, dryade à huit pétales. Dríade

SECT.: B O E P. ALT.: (1150)1700 - 2500 m. Ch. Bor.-alp. F

This plant takes its name from the dryads, the nymphs of oakwoods, as its lobed leaves, green above but white beneath, resemble those of such trees. It is an evergreen dwarf prostrate shrub that is capable of bearing the weight of the snow. It has beautiful, white flowers with at least eight petals, hence its Latin epithet. In the central-European plains a number of fossils of this plant have been found dated from the last cold period of the latest ice age —called Younger Dryas by geologists— making reference to the colonisation by this and other associated plants of the ice-free areas during that time, some 13,000 years ago.

Eco.: It colonises northern upper mountainsides, steep and calcareous, with prolonged snow cover, where it forms thick carpets on many occasions together with *Salix pyrenaica*, although it doesn't avoid the wind-swept ridges.

Loc.: Ordiso, Lapazosa ravine (B); Faja Pelay, Faja de las Flores (O); Castillo Mayor, Lapayón ravine (E); Pala de Montinier, Faja Tormosa (P).

49 Eryngium bourgatii Gouan (Umbelliferae)

Mediterranean Sea-Holly. Panicaut de Bourgat. Cardo de puerto. *Alguerrico*

SECT.: B O T V A E P. ALT.: 1300 - 2200 m. H. Plurireg. (Alp.-Med. Mount.). C

This "cardo de puerto" has bluish flowers and spiny leaves. Its compact inflorescence could be a bit misleading as this might be taken for a thistle when it actually belongs to the carrot family. Like its lowland relative, *Eryngium campestre*, it has various traditional, preventive uses to protect stalled animals and it is also used in popular medicine so as to avoid or heal skin ailments.

This blue sea-holly was first described in the 18th century from the northwestern Pyrenees by the French botanist Antoine Gouan, who dedicated it to Bourgat, a practitioner at Perpignan. It also grows in other mountains of the Iberian peninsula and northern Morocco.

Eco.: It is common in green grazed pastures of the montane and subalpine zones.

Loc.: Ordiso, Lapazosa ravine (B); Soaso, Cola de Caballo (O); Cuello Arenas, Capradiza ravine (A); La Valle (E); El Felqueral, Estiba de Espierba (P).

50 Euphorbia cyparissias L. (Euphorbiaceae)

Cypress Spurge. Euphorbe faux cyprès. Lechetrezna de puerto

SECT.: B O T V A E P. ALT.: 820 - 2040 m. H. Eur. C

Its surname indicates its resemblance to a little cypress, with numerous non-flowering branches. There exists a curious combined process affecting the pollination of this spurge and the dispersal of a particular fungus (*Uromyces pisi*). This infects the plant and transforms some of the leaves in structures very similar to the true flowers in shape, colour and odour. The fungus enters into the leaves and even produces a substance practically identical to nectar as to attract pollinators; while moving between the two types of flowers insects harvest both the nectar of true flowers and the hyphae of the fungus. Nevertheless, the fungus requires spending a short period of its life cycle in another plant, a leguminous.

Eco.: It grows in various types of montane and subalpine pastures across most of the Pyrenees, where it has not been seen west of Tena valley.

Loc.: Pastures in Otal valley (B), pastures by the parking lot, Gradas de Soaso (O), la Ripareta (A), meadowland of la Valle (E) cirque of Pineta and la Larri (P), among others.

51 **Festuca eskia** Ramond ex DC. (Poaceae)

Fétuque eskia. *Alambrón, gispet*

SECT.: B O T A E P C. ALT.: (1650)1800 - 2680 m. H. Pyr.-Cant. F

A grass with tough, pungent leaves. It forms garlands creeping down the slopes, and favours snow movement and avalanches. It is endemic to the Pyrenees and the Cantabrian mountains.

Eco.: It colonises siliceous mountainsides, in consequence it is more abundant in the peripheral zone of the Park rather than in its core, preferably on subalpine and alpine southern slopes. We can also see this plant in dense pastures together with Nard Grass (*Nardus stricta*), as well as forming steps due to slope slides (solifluction), where its roots and shoots prevent erosion.

Loc.: Tozal de las Comas, Plana Cuasta (B); Fraucata, Punta Acuta (O); Sierra Custodia, Cuello Viceto (A); Montaña de Sesa, Tozal de Basones (E); Cirque of Pineta, Añisclo Pass (P).

52 **Festuca gautieri** (Hackel) K. Richt. subsp. **scoparia** (Kerner ex Nyman) Kerguélen (Poaceae)

Spiky Fescue, Bearskin Fescue. Fétuque à balais. *Siso*

F. scoparia Kerner ex Nyman

SECT.: B O T V A E P C. ALT.: (730)1150 - 2700(2847) m.
H. Med. Mount. W CCC

Its leaves are slightly shorter and curved than those of Festuca eskia and besides its vegetative shoots are composed of bundles of curved leaves resembling a little broom, hence its Latin name (*scoparia*).

Eco.: This species is very abundant in the Park, forming large, discontinuous expanses of pastures above timberline on calcareous soil, usually in sunny spots, serving as an effective pioneer on screes and contributing to their stabilization.

Loc.: Otal, Lapazosa ravine (B); Mondarruego, Tobacor, Faja Pelay, Monte Perdido (O); Sierra Custodia, La Pardina ravine, Plana Canal (A); Tozal de la Monesma, Foratata (E); Montaspro, Estiba de Espierba (P).

53 Gentiana acaulis L. (Gentianaceae)

Stemless Gentian. Gentiane acaule. Genciana azul. *Cluchito*
G. kochiana Perr. & Song.

SECT.: B O T A E P C. ALT.: 1300 - 2415 m. H. Alp. C

This is a very showy species with big deep blue flowers compared to the rest of the plant; it has more or less lanceolate leaves with parallel margins. Despite its Latin surname (*acaulis*, stemless), the stem, at least a short one, is always visible.

However, a close relative, the Alpine Gentian, ***Gentiana alpina***, is completely stemless with short rounded leaves (small image).

Eco.: It grows in acidophilous montane or subalpine pastures.

Loc.: Espelunz, San Nicolás de Bujaruelo (B); Faja de las Flores, Faja Pela (O); Cuello Arenas, Fuen Blanca ravine (A); Cuello Viceto, Castillo Mayor (E); Cirque of Pineta, La Larri (P).

54 Gentiana verna L. subsp. verna (Gentianaceae)

Spring Gentian. Gentiane de printemps. Genciana de primavera. *Cluxitos*
Incl. **G. verna** var. **willkommiana** P. Monts. & L. Villar

SECT.: B O T V A E P C. ALT.: (1080)1320 - 2900 m. H. Bor.-alp. CC

This gentian's flower is dee[p] blue with a long, narrow tub[e] that opens in five large lobe[s] alternating with five small ap[-]pendages, resulting in that th[e] plant can be pollinated only b[y] butterflies which bear a lon[g] proboscis. The name of the ge[-]nus refers to king Gentius fro[m] Illyria (2nd century B. C.) wh[o] announced the medicinal pro[-]perties of some gentians. It takes its Latin epithet (*verna*) from it[s] bloom in early spring; however, it may have a second one in autum[n] if the climate cools and there is enough rain.

Eco.: It inhabits various types of pastures, either dense or stony, generall[y] on limey soils; it shows a wide altitudinal range, from the montane to th[e] alpine belt.

Loc.: Plana Cuasta, Bujaruelo Pass (B); Meadowland of Ordesa, Tobacc[or] (O); Cuello Gordo, Plana Canal (A); Cuello Viceto, Foratata (E); Balcón d[e] Pineta, Montaspro (P).

55 Geranium cinereum Cav. subsp. cinereum
(Geraniaceae)
Ashy Cranesbill. Géranium à feuilles cendrées. Geranio ceniciento
SECT.: B O T V A E P C. ALT.: (1450)1600 - 2800 m. H. LatePyr. CC

The leaves of this beautiful geranium grow close to the ground and are greyish, as its Latin surname suggests (cinereum, ashy); it also has showy, big, pinkish-white flowers. This plant was first described by the Valencian botanist Antonio José Cavanilles, who appeared on the reverse side of the 2000-pesetas banknote.

Eco.: It decorates stony pastures, especially those wind-swept, but also the snow-cloaked ones, preferably on limestone.

Loc.: Both the uppermost and lowermost localities of the entire Pyrenees are found within the Park. Lapazosa ravine, Col d'Oulettes (B); Cirque of Soaso, Góriz (O); Cuello Gordo, Morrón de Arrablo (A); sources of Yaga (E); trail to Collado de Añisclo, Estiba de Espierba (P).

56 Iris latifolia (Mill.) Voss (Iridaceae)
English Iris. Iris à larges feuilles. Lirio de puerto. *Allets, zebollino*
I. xiphioides Ehrh., **Xiphion latifolium** Mill.

SECT.: B O V A E P C. ALT.: 1275 - 2300 m. G. Pyr.-Cant. C

This flower is quite close to the cultivated German Iris (*I. germanica*), although the former has a bulb instead of a rhizome, and it has been used for ornamental purposes as well. It holds a moderate toxicity and in addition both species have been employed in popular medicine and also for magical rites. Its essential oil is used in some perfumes and spirits. The curious shape of its petals has been represented on a number of coats of arms, as it is one of the so-called fleur-de-lys. It is endemic to the Pyrenees and the Cantabrian Mountains.

Eco.: It blooms in midsummer and its is relatively abundant in rather fresh montane or subalpine pastures, chiefly on sheltered southern slopes.

Loc.: Ordiso, Lapazosa ravine (B); Faja Canarellos, cirque of Soaso (O); Llano Tripals, Fuen Blanca (A); Foratata, Angonés ravine (E); El Felqueral, Faja Tormosa (P).

57 Leontopodium alpinum Cass. subsp. alpinum

(Asteraceae)

Edelweiss. Edelweiss. Flor de nieve. *Flor de nieu*

L. nivale (Ten.) Huet ex Hand.-Mazz.

SECT.: B O V A E P C. ALT.: (1030)1500 - 2800 m. H. Alp. C

The edelweiss or "flor de nieve" (snow flower) is composed of a series of small capitula subtended by several special long leaves (the bracts) forming a starlike crown and covered with very dense white hairs, which gives the plant its attractive aspect. In high-mountain environments such dense cover is very useful in protecting the plant against ultraviolet radiation, cold and dehydration. The name *Leontopodium* means "lion's foot", after its downy flower head that resembles a feline paw.

Eco.: This plant is characteristic of the stony pastures on limestone, nicely spotted with its white inflorescences, from the upper montane to the alpine zone. It is more rarely found at lower altitudes, in Pineta as well as in Ordesa, the lowermost in the Pyrenees being near Sangons bridge (Añisclo), at 1,030 m.

Loc.: Bco. de Lapazosa, Batanes ravine (B); Mondarruego, Cirque de Soaso (O); Fuen Blanca, Cuello Arenas (A); Foratarruego, Sources of Yaga (E); Faja Tormosa, Estiba de Espierba (P).

58 Merendera montana (L.) Lange (Liliaceae)

Mérendéra des Pyrénées. Quitameriendas. *Espantaberaneantes, mataborregos*

M. pyrenaica (Pourr.) P. Fourn., **Colchicum montanum** L.

SECT.: B O T A E P. ALT.: 1020 - 2420 m. G. Med. Mount. W C

This bulbous plant produces violet-pink flowers that open at ground level. It has a great capability of colonisation as it uses a double strategy. Firstly, the leaves appear in autumn, after the bloom, and disappear in summer when the grazing pressure is at its peak. Besides, it deters herbivores by yielding colchicine, a toxic alkaloid that concentrates especially in its leaves.

It has also been observed that places gathering a higher density in corms (bulbs) of *Merendera* are those highly dug up by voles, and therefore some authors suggest a co-operative relationship: when voles feed on the much less toxic underground parts, they trigger a number of asexual reproduction processes which generate new bulbs, what has not been noted in undisturbed pastureland.

Eco.: It grows in various types of pastures, both the dry ones and the relatively green, from the lowland to the lower alpine zone.

Loc.: Bujaruelo Pass, Cardal ravine (B); Tobacor, Soaso, Góriz (O); Cuello Arenas, San Vicenda (A); La Loresa, Angonés ravine (E); Cinca waterfalls, La Larri (P).

59 **Narcissus moschatus** L. (Amaryllidaceae)

Daffodil. Jonquille de montagne. Narciso. *Narziso*

N. alpestris Pugsley, **N. pseudonarcissus** subsp. **moschatus (L.)** Baker

SECT.: B O T A E P. ALT.: 700 - 2160 m. G. Pyr. F

)affodils are bulbous plants that ›loom just after the snow has :hawed at the end of the winter, ›utting the first touch of colour on :he oncoming spring. This particu-ar one is ivory-coloured although :here are another daffodils within :he Park such as the Rush-Leaf Jon-quil, ***N. assoanus*** (small image), which is smaller and yellow and nuch less common (Andecastieto, =oratata). Daffodils have a highly :oxic alkaloid, the narcisine, of pa-'alyzing action.

:co.: *Narcissus moschatus* thrives in fresh, somewhat acidified pastures. \t lower altitudes It is also found on trampled forest clearings.

.oc.: Ordiso (B); Cola de Caballo (O); head and lateral ravines of Añisclo, Aolino de Aso (A); La Valle, Escuaín, Foratata (E); La Larri (P).

60 **Nardus stricta** L. (Poaceae)

Moor Mat Grass, Nard Grass. Nard raide. Cervuno. *Zerrillo, zerbuno*

SECT.: B O T V A E P. ALT.: (1355)1585 - 2620 m. H. Bor.-alp. C

This is a densely tufted grass with comb-like bluish-violet inflores-:ences where the florets all face the same direction. The sheaths ıre also characteristic, a light grey-green colour. It is scarcely ap-›reciated by cattle or livestock, which if at all harvest the new ,hoots, but they commonly reject and spit it out due to its low ›alatability.

t forms large stretches of grasslands called *tascas* above the Park :ree line, as its root system is largely developed, possessing a high :apacity of resprouting which is favoured by some mycorrhizae fungi lodged inside its roots) that help absorb nitrates.

:co.: It grows on siliceous or acidified, deep soils, very often where the ,now accumulates and also around hygroturbous pastures, mountain lakes ınd pools.

.oc.: Planabé ravine, Bujaruelo Pass (B); Aguastuertas de Carriata, Góriz O); Cuello Arenas, Plana Canal (A); Foratata, Foratarruego (E); Faja =ormosa, Añisclo Pass (P).

61 Paradisea liliastrum (L.) Bertol. (Liliaceae)
St. Bruno's Lily. Lis de saint Bruno. Flor de lis

SECT.: O P. ALT.: 1500 - 1850 m. G. Alp. F

This relative of lilies has showy white flowers, two to six of ther facing the same direction, lightly fragrant.

The genus is dedicated to count G. Paradisi.

Eco.: It lives in pastures and hay meadows a bit nitrogen-rich, on deep soi

Loc.: Soaso (O); along the trail t Añisclo Pass (P). So far, Ordes constitutes its westernmost lim in the Pyrenees.

62 Polygonum viviparum L. (Polygonaceae)
Viviparous Bistort. Renouée vivipare. Bistorta vivípara
Bistorta vivipara (L.) Delarbre

SECT.: B O A E P. ALT.: (1330)1650 - 3015 m. H. Bor.-alp. C

A small herbaceous plant with a curiou reproductive system; apart from bea ring some white or pink flowers at th top of its stem, it benefits from asexua reproduction. It produces at the sam inflorescence some reddish bulbil from which the new genetically iden tical seedlings will spring up, a syster recalling that of the viviparous animals, hence its *surname*. Man high-mountain species have alternative reproduction methods t pollination as pollinators tend to be rare in this greatly hostile en vironment.

Eco.: This plant is characteristic of snowbeds and snow-cloaked or hygrc turbous pastures, being one among those which live above the mythica elevation of 3,000 metres, to be exact at 3,015 metres in the Gabiet Central, its uppermost point in the Pyrenees. Its lowermost populations i the Aragonese Pyrenees are also found within the Park.

Loc.: El Cebollar ravine, Bujaruelo Pass (B); Soaso springs, Faja Lueng (O); Cuello Gordo, Fuen Blanca (A); Tozal de San Vicenda (E); Faja Tormc sa, Balcón de Pineta (P).

63 Primula elatior (L.) L. subsp. intricata (Godr. & Gren.) Widmer (Primulaceae)

Oxlip. Primevère intriquée. Primavera. *Flor de las neberas, primabera.*

P. intricata Godr. & Gren.

SECT.: B O A E P. ALT.: (1450)1650 - 2280(2470) m. H. Alp. F

This plant develops its rugose, undulating-margined leaves at ground level. The relatively long stem (the epithet elatior refers to this fact) springs up in the centre of each rosette bearing the flowers that bloom just after the snow thaw, spotting the pastures above the tree line with pale yellow together with *Soldanella alpina*. In popular medicine it has been used to heal wounds and is also an edible vegetable.

The Entire-Leaved Primrose ***Primula integrifolia*** (small image), with pink flowers and entire leaves is found in similar habitats up to 3,000 metres.

Eco.: It grows in dense alpine pastures especially on calcareous substratum, with prolonged snow cover.

Loc.: Lapazosa ravine, Otal (B); Faja Pelay, Arrablo Pass (O); Sierra Custodia, Sestrales (A); Canal Foratata (E); Cinca waterfalls, Faja Tormosa (P).

64 Salix pyrenaica Gouan (Salicaceae)

Pyrenean Willow. Saule des Pyrénées. Sauce del Pirineo

S. ciliata sensu DC.

SECT.: B O A E P. ALT.: (1250)1500 - 2460 m. Ch(NPdec.). Pyr. C

It is a creeping dwarf willow, the most abundant among the four to be found within the Park high mountains (the others being *S. retusa, S. reticulata, S. herbacea*). A number of fungi exclusive to these alpine willows have been found living on its stems and leaves.

Eco.: This is endemic to the Pyrenees and inhabits on calcareous stony pastures, chiefly on north-facing, snow-cloaked slopes, in the subalpine and alpine zone. However, in Ordesa (Turieto Bajo), where temperature inversion phenomena repeatedly occur, it has its lowermost location in the entire Pyrenees.

Loc.: El Cebollar, Lapazosa ravine (B); Faja de las Flores, Faja Pelay (O); Fuen Blanca (A); Tozal de San Vicenda, Castillo Mayor (E); Balcón de Pineta, Faja Tormosa, Pala de Montinier (P).

65 Saponaria caespitosa DC. (Caryophyllaceae)

Saponaire gazonnante. Jabonera cespitosa

SECT.: B O T V A E P. ALT.: 1550 - 2300 m. Ch. Pyr.-Cant. **E**

A perennial cushion-like plant which in summer is covered with pink blooms, having a stout root that penetrates deeply into the ground. The genus name comes from the Latin word meaning soapwort, as some species of Saponaria contain high concentrations of saponine which has detergent properties, whilst its surname refers to its being densely caespitose (turfing).

Eco.: It colonises wind-swept ridges and screes, from the upper montane to the subalpine belt. It is endemic to the Pyrenees and the Cantabrian Mountains.

Loc.: Tozal de las Comas, Cardal ravine (B); Sierra de las Cutas (O); Cuello Arenas, La Pardina ravine, Sestrales (A); Foratarruego (E); Cinca waterfalls, Estiba de Espierba (P).

66 Soldanella alpina L. subsp. alpina (Primulaceae)

Alpine Snowbell, Blue Moonwort. Soldanelle des Alpes. Soldanela

SECT.: B O T V A E P C. ALT.: (1400)1640 - 2600 m. H. Alp. **C**

This delicate perennial, a first cousin of primroses, blooms even before the snow has thawed, producing bell-shaped, fringed, violet flowers. The name of the genus refers to its rounded leaves that resemble an old golden coin or *soldus*.

Eco.: It is found in snowbed communities and snow-cloaked pastures on deep soil.

Loc.: Espelunz, Ordiso (B); Faja Pelay, Tobacor (O); Fuen Blanca, La Pardina ravine (A); Foratarruego, Castillo Mayor (E); Faja Tormosa, Balcón de Pineta (P).

67 Teucrium pyrenaicum L. (Lamiaceae)

Germandrée des Pyrénées. Angelitos, teucrio pirenaico.
Yerba gatera

SECT.: B O T V A E P C. ALT.: (700)965 - 2300(2450) m. Ch(H). LatePyr. **CC**

t is a perennial plant of creeping habit with a conspicuous nflorescence although its flowers are not too large. It presents two ubspecies, the typical one that has two-coloured flowers, whitish ınd purple (big image), and subsp. ***guarensis*** P. Monts. with white corolla (small image). It has been used in popular medicine as an nfusion or tea, to heal colds and it also facilitates the expulsion of idney stones.

co.: It grows in stony dry pastures, scrub, forest fringes and mountain crees, chiefly on calcareous soil, from the lower montane to the upper lpine zone. It is unique to the Pyrenees and the Iberian System.

oc.: Ordiso, Planabé ravine (B); radas de Soaso, Carriata (O); an Úrbez, La Pardina ravine (A); evilla, La Valle (E); Montaspro, irque of Pineta (P).

68 Trifolium alpinum L. (Fabaceae)

Alpine Clover. Trèfle alpin. Regaliz de puerto. *Mandilo*

SECT.: B O T V A E P C. ALT.: (1350)1600 - 2600 m. H. Alp. **CC**

This clover has one of the largest inflorescences among those which grow in our mountains, with intense pink flowers. It has a high nutricional value and so it is much appreciated by livestock, forming real meadows. Its stout root can exceed one metre in length and contains glycyrrhicine, the ame active ingredient as Liquorice (*Glycyrrhiza glabra*), and for his reason it is used in popular medicine to heal digestive ulcers, nflamation, respiratory and dermatological ailments.

co.: It colonises high-mountain dense pastures on deep soil, rich in rganic matter and more or less acidified. It is found from the Cantabrian Aountains to the Carpathians.

oc.: Bujaruelo Pass, Ordiso (B); Góriz, Millaris (O); Sierra Custodia, Llano ripals, (A); Montaña de Sesa, Foratarruego (E); La Larri (P).

HAY MEADOWS

Hay meadows are semi-natural ecosystems, that is to say they are influenced by human actions and depending on its management (number of cuts per year, irrigation, fertilization, grazing and even sowing) they can be home to different floristic elements. In general they derive from former panares (mountain cereal fields, normally of rye or wheat) or potato crops, transformed into meadows intended for hay production to feed the livestock, mainly cattle, during wintertime. In some cases, nature can, after the initial sowing, interact with human management, while in others, the meadows are re-sown regularly.

Nowadays the Pyrenean meadowlands are suffering a serious setback as many locals have stopped farming to work in the tourism industry or retire, and a number of meadows become residential areas. Abandonment or extensive management transforms them into dry or semi-dry pastures in the foothills, invaded by gorse (*Genista scorpius*), hawthorn (*Crataegus monogyna*) and blackthorn (*Prunus spinosa*) as is the case with Añisclo (San Úrbez-Bordas de Aso); or they evolve towards oak-pine woodland, as we can see in Escuaín and Añisclo; in the fresher areas they become more humid pastures (Bordas de Salarons, Casa Oliván, Laña Caballo in Ordesa, or La Valle in Escuaín). Near the streams or in particularly fresh spots of the mountainsides they result in hygrophilous pastures with tall plants such as *Heracleum sphondylium* subsp. *pyrenaicum* (Soaso in Ordesa or at Fuen Blanca and Ripareta in Añisclo).

These days the only functional meadows within the Park are in Andecastieto in Ordesa, just by the old Parador visitor's centre. However, in the periphery of the protected reserve we can see fine examples of a still productive meadowland in Broto, Puértolas-Bestué and Pineta valleys.

69 **Arrhenatherum elatius** (L.) Beauv. ex J. & C. Presl. subsp. **elatius** (Poaceae)

False Oat-grass. Fenasse. Avena elevada. *Feno, yerba formental*

Avena elatior L.

SECT.: B O T A P C. ALT.: 940 - 2150 m. H. Plurireg. F

A tall plant, as its Latin epithet *elatius* indicates, and one of the grasses sown in the meadows to be a part of hay bales.

Eco.: It thrives in hay meadows forming loose turfs, and can also be found in other herbaceous communities, as well as on roadsides, forest clearings, etc.

Loc.: L'Artica ravine, Planabé (B); Andecastieto, Bordas ("barns") de Salarons (O); San Úrbez, Fuen Blanca (A); Llanos de Diera, El Plan (P).

70 **Astrantia major** L. subsp. **major** (Umbelliferae)

Great Masterwort. Grande Astrance. Astrancia

SECT.: B O T A E P. ALT.: 920 - 1960 m. H. Eur. F

The great masterwort bears tiny greenish-white or light pink flowers that stand out because they are clumped together in an inflorescence ringed with attractive greenish-white leaf-looking bracts.

Eco.: It grows in fresh forests, both deciduous and evergreen, and also in megaphorbs and hay meadows, always on fertile calcareous soils.

Loc.: Faixa O Cordiador, Oncins bridge (B); Casa Oliván, Senda de Cazadores (O); Airés ravine, Añisclo canyon (A); Tabacoy, Yaga gorge (E); Tormosa ravine, El Felqueral (P).

71 Chaerophyllum aureum L. (Umbelliferae)

Golden Chervil. Cerfeuil d'or

Ch. maculatum Willd. ex DC

SECT.: B O T A E P C. ALT.: 950 - 2100 m. H. Alp. **E**

t is a slender mountain plant with divided triangular leaves. The epithet aureum refers to the yellowish or golden colour of the fruit.

Eco.: It thrives in hay meadows and herbaceous communities with soil rich in organic matter and nitrogen, even in megaphorbs, always in humid, broadleaved-forest environments.

Loc.: Sandaruelo ravine (B); Gradas de Soaso, at the foot of Tozal del Mallo (O); Sestrales, Añisclo canyon (A); Revilla (E); Parador, trail to the Waterfalls (P).

72 Crepis biennis L. (Asteraceae)

Rough Hawksbeard. Crépide bisannuelle

SECT.: T P. ALT.: 1060 - 1450 m. H. Eur. **RR**

This is a fairly robust plant up to one metre in height. As its surname indicates, it needs two years to complete its life-cycle. In Spain it is relatively rare, its westernmost localities of the Aragonese Pyrenees being in the Park periphery.

Eco.: It is found in humid, fertilized hay meadows. The progressive disappearance of many of them might be the reason for its scarcity these days.

Loc.: next to the hermitage of Sta. Ana de Diazas (T); El Plan, Javierre meadowland (P).

73 Geranium sylvaticum L. subsp. sylvaticum

(Geraniaceae)

Wood Cranesbill, Woodland Geranium. Géranium des bois. Geranio de bosque

SECT.: B O E P. ALT.: 1170 - 2070 m. H. Bor.-alp.

The woodland geranium (hence its epithet *sylvaticum*) has larg violet-purple flowers assembled in pairs.

Eco.: It inhabits megaforbs and humid herbaceous communities in fores clearings, particularly those rich in organic matter, as well as place higher up frequented by cattle.

Loc.: Santa Elena bridge, Las Comas ravine (B); Cotatuero, Senda de lo Cazadores (O); Castillo Mayor (E); El Felqueral, La Larri (P).

74 Heracleum sphondylium L. subsp. pyrenaicum

(Lam.) Bonnier & Layens (Umbelliferae)

Common Hogweed. Berce de Pollini. Belleraca. *Cornases, pampas*

H. pyrenaicum Lam.

SECT.: B O A E P. ALT.: 1100 - 2180 m. H. Mount. N Iber.-Pen.

A stout plant with large and deepl divided leaves, loathed by farmer as its stem is quite thick and hard t cut so that it causes compacting pro blems in hay bales; therefore the intend to wipe it out of the meadow by all means possible, for instanc pouring salt inside the hollow stem The genus name of this plant, *He racleum*, refers to its having bee thought so strong that it was compared to the strength of Heracle or Hercules, the Greek hero.

Eco.: It grows in herbaceous communities on deep or stony soils, rich i organic matter. It is usually found at the foot of shady rocks and in clea rings of humid woodland, or in places frequented by the livestock, etc. also invades mismanaged or unused hay meadows.

Loc.: Faixa O Cordiador, El Turbón (B); Góriz, Gradas de Soaso (O); San Úrbez Fuen Blanca (A); Los Mallos bridge (E); Parador, Las Inglatas (P).

75 Rhinanthus pumilus (Sterneck) Pau subsp. pumilus

(Scrophulariaceae)

Rattle. Rhinanthe du Midi. Cresta de gallo. *Cascabelera*

Rh. mediterraneus (Sterneck) Sennen

SECT.: B O T V A E P. ALT.: 940 - 2350 m. Th. Latesubmed. C

This is an annual hemiparasitic plant. Its flowers have a swollen calyx that might resemble a small bell whilst the yellow corolla is nose-shaped (hence the name of the genus), ending in a little violet tooth more than one millimetre long. When it fruits, the seeds make a rattling noise if the plant is shaken.

Eco.: It is frequently found in hay meadows and mesophilous grasslands of the montane and subalpine zone.

Loc.: Duáscaro, Turbón (B); Andecastieto, Soaso cirque (O); Cuello Arenas, Fuen Blanca (A); Puértolas (E); Las Inglatas, Tormosa ravine (P).

76 Trisetum flavescens (L.) Beauv. (Poaceae)

Golden Oat Grass, Yellow Oat Grass. Avoine dorée. Avena rubia

Trisetaria flavescens (L.) Baumbg.

SECT.: B O T V A E P. ALT.: 940 - 2300 m. H. Lateeur. E

It is a grass that can grow more than one metre height. As the image shows, its flowers are grouped in a loose, delicate inflorescence —actually a panicle— up to 20 cm long, which, depending on the sunshine, gives off bright yellowish-green hues, hence its *surname*.

Eco.: It is one of the most abundant plants growing in hay meadows, as well as in grasslands, humid pastures and forest clearings, on deep, fertile soils.

Loc.: Sandaluelo and Lapazosa ravines (B); Andecastieto, halfway between the refuge of Góriz and Faja Luenga (O); Diazas (T); Sestrales, Forquela (A); refuge of Foratata, Castillo Mayor (E); El Plan (P).

WETLANDS

TThe rocky substratum in the Park is mostly calcareous in origin, so the rocks are very permeable and therefore the formation of wetlands is unlikely. In Ordesa and Monte Perdido such places are rare and highly localised, chiefly in drains and in the so-called «aguastuertas» or «catuartas» (meandering streams) and in springs whose presence indicates a geological contact of different permeability.

The true central-European peatlands with *Sphagnum* are not really formed here, but hygroturbous pastures whose waters have a neutral-alkaline pH. This singularity has enabled the Park to house a plant community unique and endemic to this central section of our mountain range, a relative of those to be found in the Alps, which might have entered the Pyrenees during the last ice ages. Outside the Park limits, at the source of the Ara river where the substratum is granite or schist, true peatlands exist.

77 **Carex bicolor** All. (Cyperaceae)

Two-colour Sedge. Laiche bicolore. Cárice bicolor.

SECT.: O A. ALT.: 2155 - 2430 m. H. Bor.-alp. F

This is a tiny herb, very rare in the Pyrenees, having most of its populations within the Park.

Its name refers to the two coloured trait of the ripe spikes, with a contrast between the bluish-green of the utricles and the brownish-grey of the bracts.

Eco.: It grows in the alpine belt, in damp spots on limestone substrate with a little incline which generate hygroturbous grasslands unique to the central Pyrenees.

Loc.: Faja Luenga, Faja Millaris and Aguas Tuertas at Salarons (O); Morrón de Arrablo (A).

78 **Carex davalliana** Sm. (Cyperaceae)

Davall's Sedge. Laiche de Davall. Cárice de Davall

SECT.: B O T A E P. ALT.: (1260)1645 - 2330(2680) m. H. Eur.

An easily recognizable sedge whe it flowers and bears fruit as it is dioecious plant which produces tw types of turfs, some male (righ image), some female (left image).

It is dedicated to the English bota nist Edmund Davall, who worked i Switzerland during the 18th century

Eco.: This herbaceous plant form tussocks in springs and wetlands tha grow on calcareous soil in the uppe montane and subalpine belts, the so called hygroturbous meadows.

Loc.: Oncins spring, Batanes ravin (B); Briet and Soaso springs (O) Sierr Custodia, Fuen Blanca (A); Faja Mones ma (E); Las Inglatas, Faja Tormosa (P).

79 Cirsium monspessulanum (L.) Hill
subsp. **monspessulanum** (Asteraceae)
Cirse de Montpellier. Cardo de Montpellier

SECT.: B O T A E. ALT.: (600)1000 - 1650(1800) m. H. Med. W E

This fairly tall thistle that flowers in midsummer has slightly fleshy leaves -the cauline ones embracing the stem- and soft slender spines. The name refers to the region of Montpellier, in France, the place of origin of the plants that served to its original description.

Eco.: It is a characteristic species of the rush areas that grow on tufa springs, wet meadows, etc.

Loc.: Springs of Santa Elena and El Azute (B); La Canal ravine, Diazas (O); San Úrbez, Puyarruego (A); ravines of La Consusa and Angonés (E).

80 Epipactis palustris (L.) Crantz (Orchidaceae)
Marsh Helleborine. Épipactis à longues feuilles. Epipactis de fuentes

SECT.: O A E P. ALT.: 600 - 1300 m. G. Lateeur. R

This orchid differs from the rest of its fellow relatives in its large white labellum and the greenish-purple sepals. It is pollinated either by ants or bumblebees.

Eco.: It grows in wetlands, calcareous springs or by the rivulets of the lower montane zone, hence its *surname*.

Loc.: We can enjoy its beautiful bloom at the end of June in the marshes close to Casa Oliván (O), also in Añisclo gorge or in the wet meadows opposite the former Sanatorio de Pineta, as well as near Escuaín.

81 Eriophorum latifolium Hoppe (Cyperaceae)

Cottongrass, cottonsedge. Linaigrette à larges feuilles. Hierba algodonera. *Yerba algodonera*

SECT.: B O A E. ALT.: (1300)1670 - 1900(2330) m.
H. Plurireg. (Bor.-alp. y Eur.). R

This species bears several spiklets with a scabrid peduncle whilst its seeds are crowned with a white cotton-like plume to aid dispersal, the word *Eriophorum* meaning that it "wears wool".

Eco.: This plant lives exclusively in hygroturbous pastures, springs and pool shores, chiefly calcareous, from the montane to the lower subalpine zone.

Loc.: Oncins spring (B); Faja Pelay, Tobacor, Briet's spring (O) Custodia, above La Capradiza (A); Faja Monesma (E).

Within the Park is also found another species of this genus, the Scheuchzer's cottongrass or white cottongrass, ***E. scheuchzeri*** (small image). It has a distinctive lone spiklet having no stalk. It is rarer than the previous one and is to be found in the same habitat, although in the alpine belt. Vilá ravine (B); Aguas Tuertas at Salarons, Faja Luenga (O).

82 Mentha longifolia (L.) Huds. (Lamiaceae)

Horse Mint. Menthe à feuilles longues. Hierba buena hedionda. *Ixalarga.*

M. spicata L. var. **longifolia** L., **M. sylvestris** L.

SECT.: B O T A E. ALT.: (600)950 - 1800(1950) m. H. Eur. E

Among the mints this one has the strongest smell to the point of being particularly unpleasant. The leaves are greyish-green on its underside and also longer than those of other mints, as its Latin name indicates. Like other species of the same sort, it has been widely used in herbal medicine and pharmacy due to its richness in menthol, an alcohol utilised as flavouring and taste corrector, as well as for its multiple therapeutic properties. According to the Greek myth, Minthe, the daughter of Kokytos, was changed into this plant by Proserpina.

Eco.: It is associated with rush areas and herbaceous communities in springs or streamsides used by cattle as drinking spots, and therefore nitrified.

Loc.: Oncins spring, Labaza ravine (B); Briet's bridge, La Vaqueriza (O); San Úrbez, La Mirona ravine (A); Foratata, La Garganta ravine (E).

33 Molinia caerulea (L.) Moench (Poaceae)

Purple Moor-Grass. Molinie bleue. Escoba. *Escobizo, lastón d'escobas*

SECT.: B O T A E P C. ALT.: (600)950 - 2100 m. H. Lateeur. F

This is a perennial tussock-forming grass with stout rhizomes from which new stems, more than one metre high, emerge. It flowers at the end of the summer producing a long bluish-green spike (hence its epithet *caerulea*). It has decidous tapering leaves which stay in the plant turning light grey-brown until the next season, a fact that reveals the presence of springs from afar. The genus is dedicated to J.I. Molina, a Jesuit and naturalist settled in Chile in the 18th and the 19th centuries.

Eco.: It grows in travertine springs, seeps and wet grasslands.

Loc.: Sandaruelo ravine, Otal (B); Briet's springs, Cotatuero (O); Selba Plana, Fuen Blanca (A); Faixa Billa, Angonés ravine (E); Las Inglatas, El Felqueral (P).

34 Parnassia palustris L. subsp. palustris (Saxifragaceae)

Marsh Grass-of-Parnassus, Bog-Star. Parnassie des marais. Hepática blanca

SECT.: B O T A E P C. ALT.: 950 - 2350 m. H. Bor.-alp. C

A little plant bearing white flowers with an ingenious pollination system. Each flower has 5 stamens that ripen alternatingly -curving themselves inwards-; moreover, it has a hardly productive nectary from which a series of filaments protrudes. Each filament is topped with a gleaming cap that leads the insects to believe that there is some nectar on them. So when they visit the flower and try unsuccessfully to get some food, they rub themselves against the stamens or the stigma and so pollinate the plant. This takes its name from a herb that, according to the classical botanists, use to grow in Mount Parnassus, in Greece, the home of the Muses.

Eco.: It inhabits springs as well as wet, hygroturbous pastureland, mainly on limestone areas.

Loc.: Planabé, Santa Elena (B); Aguas Tuertas at Carriata, La Canal ravine (O); San Úrbez, La Pardina ravine (A); Faja Tormosa, Cirque of Pineta (P).

85 Petasites paradoxus (Retz.) Baumg. (Asteraceae)

Paradoxal Butterbur. Pétasite blanc de neige. Petasites.
P. niveus (Vill.) Baumg.

SECT.: A E P C. ALT.: 1230 - 1800(2050) m. G. Alp.

It is a curious herbaceous plant tha flowers very early, before the leaves ap pear; these are triangular in shape an covered with a dense fuzz underneatl Although other species of *Petasites* hav healing properties and are used in pc pular medicine (especially *P. officinalis*) uses of this sort have not been recorde in the Pyrenees, maybe due to its scarc ty. It is endemic to the European mour tains, from the Carpathians to the Pyre nees, where it appears here and there In Navarra it is protected because of it scarcity.

Eco.: It colonises humid stony spots, such as colluvial deposits and alluvi gravel.

Loc.: It has been found in Añisclo, although it becomes much mor noticeable on the Cinca riverbanks in Pineta. La Ripareta, La Pardin ravine (A); La Garganta ravine (E); Es Sacos ravine, track to La Larri (P).

86 Pinguicula grandiflora Lam. subsp. grandiflora (Lentibulariaceae)

Long-Flowered Butterwort. Grassette à grandes fleurs. Grasilla. *Biola d'aigua.*

SECT.: B O T A E P C. ALT.: (750)950 - 2490 m. H. Atl.

Pinguicula are passive carnivorous plants as they don't move to seize the small arthropods they feed on. The surface of their yellowish-green leaves is covered with sticky glands which they use to capture prey, which are attracted both by its fungal odour and the refraction of the light in the little drops of mucilage. Another species to be found in these wet environments, the Common Butterwort, ***P. vulgaris*** (upper small image), differs in having a smaller, more closed flower. There is yet one more species in the Park, the Alpine Butterwort ***P. alpina*** (white flower).

Eco.: It grows in wet pastures and seeps, chiefly on limestone.

Loc.: Soaso, Tamborrotera (O); Sangons bridge, La Abetosa ravine (A); Faja Monesma, Surgencia del Yaga (E); Faja Tormosa, Parador (P).

87 Primula farinosa L. (Primulaceae)

Bird's-eye Primrose. Primevère farineuse. Primavera de fuentes.

SECT.: B O T A E P C. ALT.: (1250)1600 - 2680 m. H. Bor.-alp. C

This little primrose turns the springs pink with its clumps of flowers. The name refers to the floury appearance of the underside of its leaves, which are arranged in a basal rosette.

Eco.: It is a plant characteristic of springs, rivulets and wetlands in the subalpine and alpine levels.

Loc.: Surely enough, it is because of temperature inversion that it has abyssal populations in the montane level at Ordesa. Briet bridge, Faja Luenga (O); Fuen Blanca, Plana Canal (A); Faja Monesma, La Valle (E); Faja Tormosa, Las Inglatas (P).

88 Saxifraga praetermissa D.A. Webb (Saxifragaceae)

Neglected Saxifrage. Saxifrage négligée. Saxífraga de ventisquero

S. ajugifolia auct. non L.

SECT.: B O A E P C. ALT.: 1750 - 3010 m. Ch. Pyr.-Cant. C

Unlike other saxifrages, this one has lateral, leafy floriferous stems. Its name means "forgotten, neglected or missing" referring to the fact that it was only a short time ago (in 1963) when it was properly described to science despite having been known years ago.

Another saxifrage also found in wet spots and seeps is the Yellow Mountain Saxifrage or Yellow Saxifrage, ***S. aizoides*** (small image).

Eco.: It colonises stony, late snow patches, wet screes and humid limestone crevices, as well as mountain rivulets, etc., from the subalpine up to the subnival level.

Loc.: An endemic species to the Pyrenees and the Cantabrian Mountains that can be found at: Forqueta del Gabieto, Batanes ravine (B); Cilindro de Marboré, Faja Luenga (O); Punta de las Olas, Bellós springs (A); Faja Monesma, Angonés ravine (E); Lake Marboré, trail to Añisclo Pass (P).

ROCKY GROUND

Massive cliffs are one the main features of this National Park where the highest limestone massif in Europe is located, with the Monte Perdido standing as its emblem. It has the broadest altitudinal range in the Pyrenees, as well as a variety of rock faces from the foothills at 700 m to the subnival zone, above 3,000 metres.

These are inhospitable habitats, only suitable for those species adapted to very restrictive conditions of soil and water, places that challenge their adaptive capacity and which, together with the screes, originate most of the plants endemic to our mountains, as we will see next.

89 Adiantum capillus-veneris L. (Adiantaceae)

Southern Maidenhair Fern, Black Maidenhair Fern, Venus Hair Fern. Capillaire. Culantrillo de pozo. *Yerba meadera*

SECT.: T A E. ALT.: 720 - 1240 m. H. Latemed. F

The name of this fern means "no wet", as the particular arrangement of its fronds lets the wate drip without soaking them up, hence its vernacular Aragonese nam "hierba meadera" (peeing herb) The name of the species, *capillus veneris*, means Venus' hair, because the rachis of the leaflets resemble the goddess's hair. In the Alt Aragón (province of Huesca) it ha been used in folk medicine eithe as a diuretic or abortive.

Eco.: This plant carpets warm cliff seeps and shady crevices in the lowe montane zone, just where limestone travertine is formed due to carbonate precipitation.

Loc.: Within the Park, the black maiden fern is only found in the warmes valleys, Añisclo and Escuain: San Úrbez cave, Fuente del Baño (A); Yag gorge(E).

90 Androsace cylindrica DC. subsp. cylindrica (Primulaceae)

Rock-Jasmine. Androsace cylindrique. Androsace columnar

SECT.: O A E. ALT.: (1500)1700 - 2385 m. Ch. Pyr. E

In the older specimens of this cushion-forming perennial we ca see the short, tightly packed, columnar little stems —that retai the remains of dead leaves— overlapping, hence the name of thi species.

Eco.: It is a specialist in growing in the fissures of limestone rock faces either vertical or overhanging, both sunny and shady, of the subalpine an lower alpine belts.

Loc.: It is endemic to the Aragonese Pyrenees, its westernmost limit bein Ordesa, where, in addition, its largest populations are found. Solan del Gallinero, Fajas Blanquera and de Las Flores, Faja Pelay (O); In th Escuain section it is only found at Castillo Mayor.

There is another very similar plant in the Park, also endemic to the Pyrenees, the Pyrenean Rock-Jasmine, *A. pyrenaica* Lam. (small image). Within the Park it has only been found at Sestrales (Añisclo) where it grows on quartzite (a siliceous rock) and has there its westernmost population.

91 Antirrhinum sempervirens Lapeyr. subsp. sempervirens (Scrophulariaceae)

Silver Snapdragon. Muflier sempervirent. Boca de dragón siempreverde

SECT.: O V A P C. ALT.: 980 - 2350(2500) m. Ch. Pyr. F

It is an evergreen perennial, as its scientific name implies, whose stems are woody at the base. This plant has a dull green colour because of being covered with a smooth fur. Its flowers are white, tinged with yellow or purple.

Eco.: It can be spotted, in the form of shrubs or isolated specimens, growing on limestone rocks, particularly those facing south.

Loc.: Another plant endemic to the Pyrenees, with its best populations located in Huesca. Faja Racón, Carriata (O); Mondoto, Betosa ravine (A); Faja Tormosa, trail to Balcón de Pineta (P).

92 Campanula cochleariifolia Lam. (Campanulaceae)

Fairies' Thimbles. Campanule à feuilles de cochléaire. Campanillas. *Campanetas*

C. pusilla Haenke

SECT.: B O V A E P C. ALT.: (1140)1400 - 2850(3071) m. H. Alp. C

The blooms of this bellflower, quite large as compared to the plant size, are light blue and shaped like the megaphone of a phonograph. Its basal leaves resemble a teaspoon, hence its Latin name, and unlike other species of the same genus, it still keeps them at flowering time.

Eco.: It takes root in screes and rock crevices, often moist, at times in stony pastureland as well, preferably on calcareous substratum.

Loc.: San Nicolás de Bujaruelo, Otal (B); Faja Pelay, Góriz ravine (O); Capradiza ravine, Liana Mala (A); Angonés ravine, Pleta de los Faixins (E); Cinca waterfalls, Eastern Oriental (P).

Also we found in Pineta Valley (small image) another endemic plant to the Pyrenees, ***C. jaubertiana*** Timb.-Lagr. (= *C. andorrana* Braun-Blanq.).

93 Chelidonium majus L. (Papaveraceae)

Tetterwort, Greater Celandine. Grande Chélidoine. Celedonia mayor. *Yerba berruguera*

SECT.: T V A E P. ALT.: 950 - 1385 m. H. Lateeur. R

A well-known plant that has yellow sap, which despite this being caustic and irritant, has been used in popular medicine to *burn* warts, among other therapeutical usages.

The Latin name of the celandine means "little swallow" as it blooms when these birds arrive.

Eco.: It grows on walls of old houses, slab roofs, and vegetable-patch stone fences.

Loc.: Molino de Aso, Gallisué (A); Revilla, Escuaín (E); Pineta hermitage (P).

94 Cystopteris fragilis (L.) Bernh. (Athyriaceae)

Brittle Bladderfern, Common Fragile Fern. Capillaire blanc. Culantrillo blanco. *Mermasangre*

C. filix-fragilis (L.) Borbás

SECT.: B O T A E P C. ALT.: 950 - 3200 m. H. Lateeur. [Subcosm.]. C

The name of the genus in this fern makes reference to the indusium, in which the spores are hosted, shaped like a small bladder (cistos). It has been used in popular medicine, either by itself or combined with other ferns such as *Asplenium trichomanes*, as a *mermasangre* (to lower blood pressure).

Eco.: This fragile fern is rooted in rock fissures, both calcareous and acidic, appearing from the lower montane up to the subnival zone.

Loc.: Marboré Peak (3,200 m) constitutes so far the uppermost place where this fern grows. Lake Bernatuara, Cerbillona ravine (B); Mondarruego, Góriz ravine (O); Mondoto, Pico Navarro (A); Montaña de Sesa, Foratarruego (E); Lake Marboré, Faja Tormosa (P).

95 Erinus alpinus L. subsp. alpinus (Scrophulariaceae)

Fairy Foxglove, Starflower, Alpine Balsam, Liver Balsam.
Érine des Alpes. Erino

SECT.: B O T V A E P C. ALT.: 700 - 2560 m. H. Plurireg. (Alp.-Med. mont). C

It is a little plant, somewhat lignified at the base, which produces a rosette of hairy leaves, from which emerge the leafy flowering stems. Its flowers are small and are grouped in racemes, with the pinkish corolla formed by a basal tube and five lobes, two upper and three lower.

Eco.: It is one of the most conspicuous plants in our flora, growing on rocks (lithophytes), as it can be found from the foothills up to lower alpine belt, chiefly on limestone.

Loc.: Gabieto ravine, Otal (B); Turieto Bajo, Faja Canariellos (O); Molino de Aso, La Pardina ravine (A); Angonés ravine, Foratarruego (E); Las Inglasas ravine, Parador (P).

96 Jasonia glutinosa (L.) DC. (Asteraceae)

Jasonie. Té de Aragón. *Té de roca*
Chiliadenus glutinosus (L.) Fourr., **Ch. saxatilis** (Lam.) Brullo, **J. saxatilis** (Lam.) Guss.

SECT.: O T A E P. ALT.: 750 - 1555(1850) m. Ch. Med. W E

This plant is woody at the base and produces leafy stems, glandulous and tacky to the touch as the name of the species shows, besides being branched at the end and topped with small yellow flower heads. This is the genuine té de roca (rock tea) that grows at noteworthy elevations in warm places such as the south face of Gallinero in Ordesa. It is widely used in popular medicine, mainly in teas, to heal digestive ailments.

Eco.: It colonises calcareous rock fissures and ledges in dry, sunny spots.

Loc.: Torla, San Antón hermitage, Faja Racón (O); San Urbez, A Liana (A); Consusa ravine, La Loresa (E).

97 Petrocoptis crassifolia Rouy (Caryophyllaceae)

Rompepiedras

P. albaredae (P. Monts.) P. Monts. subsp. **guinochetii** (J.M. Monts.) J.M. Monts.

SECT.: V A E. ALT.: 700 - 2100 m. Ch. Pyr. C. E

It is a *stone-breaker* plant as the name of the genus indicates. It produces hanging, curved stems, with opposite, somewhat fleshy leaves (hence its epithet *crassifolia* meaning thick, succulent-leaved), a matte glaucous green. This species was a subject of study by the researcher at the Instituto Pirenaico de Ecología Begoña García, who discovered that its flowers present different ultraviolet patterns to attract pollinators, when it fruits, the stems twist and face the rock as to try putting the seeds into any accessible crevice.

Eco.: It hangs, literally, on the limestone cliffs and is a specialist in colonising overhanging rock faces.

Loc.: It is an endemic plant to the Central Pyrenees whose most abundant populations live within the Park. Cueva de las Estalactitas, San Úrbez (A) Angonés ravine (E).

98 Pinguicula longifolia Ramond ex DC.

subsp. **longifolia** (Lentibulariaceae)

Long-leaved Butterwort. Grassette à feuilles longues. Atrapamoscas de hoja larga, grasilla

SECT.: B O A E P. ALT.: 650 - 2000 m. H. Pyr. C. F

This carnivorous plant has long, pale green leaves (as its Latin name indicates) of up to 20 cm, sticky and hanging, which serve to capture the arthropods that complement its diet. Among the glands on its leaves, Dr. Begoña García discovered a new subespecies of a mite (*Oribatula tibialis*, small image) which feeds on the remains of the little creatures that the plant is unable to digest, acting as a sweeper.

Eco.: It thrives on seeps of vertical or overhanging rock faces, without direct exposure to sunlight, along with other species such as *Carex brachystachys* and *Hypericum nummularium*.

Loc.: This is a remarkable endemic to the Central Pyrenees which has its westernmost distribution limit in Bujaruelo, whilst its best populations are found in the Añisclo gorge. Gorga de Santa Elena, Fuen del Azute (B) Cueva Frachinal, Faja Canariellos (O); Las Cambras, La Pardina ravine (A) Castillo Mayor, Surgencia del Yaga (E); Cinca waterfalls, Tormosa cave (P)

99 Potentilla alchimilloides Lapeyr. (Rosaceae)

Alchemilla-leaved Cinquefoil. Potentille fausse alchémille.
Falsa alchemilla
P. alchemilloides Lapeyr.
SECT.: B O V A E P C. ALT.: (965)1500 - 2485 m. H. LatePyr. C

Perennial, with silky stems and palm-shaped leaves silvery underneath. The name *Potentilla* refers to the strong, alleged properties, tonic and astringent, that some species of this genus have, whilst its surname highlights the resemblance that this plant bears to those of the *Alchemilla alpina* group, from which it differs in having very apparent white flowers.

Eco.: It colonises the fissures and ledges of limestone cliffs no matter their orientation, although at higher elevations it prefers the south facing slopes. It co-exists with *Asperula hirta*, *Lonicera pyrenaica* and other species characteristic of rocky ground forming a plant community very well represented within the Park.

Loc.: Picamartillo, Ordiso (B); Mondoto, La Pardina ravine (A); Cola de Caballo, Tozal del Mallo (O); Forata-rruego, La Garganta ravine (E); Cinca waterfalls, Faja Tormosa (P).

On the foothills of Añisclo, another plant, *P. caulescens* L. (small image), grows on limestone rock faces, as well.

100 Ramonda myconi (L.) Rchb. (Gesneriaceae)

Pyrenean Violet, Rosette Mullein. Ramondie des Pyrénées.
Oreja de oso. *Borraina, yerba cerruda*
R. pyrenaica Pers.
SECT.: B O T V A E P C. ALT.: 605 - 2350 m. H. LatePyr. C

The genus is dedicated to the French naturalist of the 18th century Louis Ramond de Carbonnières, who pioneered the botanical exploration of Monte Perdido massif. After a period of drought, this plant seems to come back to life, being one of the few vascular plants capable of breaking out after its total desiccation, thanks to the presence of raffinose, a carbohydrate, and the accumulation of another one, saccharose, at the moment of dehydration, which stop cell death. There is only one other plant genus in Europe with the same capability, *Haberlea*, that belongs to the same family. Maybe this feature has allowed them to survive the ice ages, as almost all the species in the *Gesneriaceae* live within the tropics.

Eco.: This endemic to the Pyrenees thrives on shady, rather humid limestone rocky ground, which it very often carpets, from the foothills to the lower alpine belt, usually in the company of *Saxifraga longifolia*.

Loc.: Santa Elena bridge, Picamartillo (B); Frachinal cave, Turieto (O); Molino de Aso, Fuen Blanca (A); Los Mallos spring, La Valle (E); El Felqueral, Las Inglatas (P).

101 Sarcocapnos enneaphylla (L.) DC. (Papaveraceae)

Fumeterre à neuf folioles. Hierba de Lucía. *Tabaco de Simó*
Fumaria enneaphylla L.

SECT.: A E. ALT.: 730 - 1500(1895) m. Ch. Med. W E

The name of the genus comes from a Greek word meaning "fleshy *Fumaria*", as its leaves resemble those of the fumeworts. The Latin epithet refers to its leaves split in nine leaflets, although this is not always the case.

Eco.: A heat-loving chasmophyte that colonises perpendicular or overhanging cliffs, preferring the south-facing ones at high elevations. At lower altitudes it lives in the company of *Petrocoptis crassifolia* or the small fern *Asplenium trichomanes* subsp. *csikii*.

Loc.: It enters into Añisclo through the Cinca valley. It is also found in Sestrales (at 1895 m), where it reaches its uppermost location in the Aragonese Pyrenees. San Úrbez, La Pardina ravine (A); San Lorién hermitage (E).

102 Saxifraga aretioides Lapeyr. (Saxifragaceae)

Saxifrage de Burser. Rompepiedras amarillo

SECT.: O A E P C. ALT.: (1200)1550 - 2230(2340) m. Ch. Pyr. E

The name of the genus, *Saxifraga*, means "stone-breaker". Perennial plant that forms rosettes of small, stout leaves, which are covered with glands (hydathodes) which secrete calcium carbonate. The stems bear narrow, closely-packed little leaves and are topped with yellow flowers. Also we found ***S. caesia*** L., with white flowers (right image).

Eco.: It grows in the fissures of limestone rocks, preferably on sunny exposures.

Loc.: It is endemic to the Pyrenees, ranging from Navarra to Lleida. Within the Park some abyssal locations are found either in Ordesa (1400 m) or in Pineta (1200 m), the latter being the lowermost of the southern Pyrenees, very likely because of temperature inversion. Senda de Cazadores, Faja de las Flores (O); Fuen Blanca, Capradiza ravine (A); Castillo Mayor (E); Estiba de Espierba, near Añisclo Pass (P).

103 **Saxifraga longifolia** Lapeyr. subsp. **longifolia** (Saxifragaceae)

Saxifrage à feuilles longues. Corona de rey. *Abortacrabas, coda de borrega*

SECT.: B O T A E P C. ALT.: 720 - 2485 m. H. Med. Mount. W CC

'his perennial can live for up to 30 years ır even longer, forming solitary long-lea-ved rosettes (as its Latin name suggests). urthermore, the leaves are narrow and covered of calcium carbonate secretions. t blooms only once in a lifetime produ-cing long, dense, pyramidal plumes with nore than 500 white flowers, and dies af-erwards. It is a relic of the Tertiary flora which apparently origina-ed in the Pyrenees and then extended as far south as to North Afri-a. It is a toxic plant of abortive properties, that was mainly used in olk veterinary medicine, as some of its vernacular names indicate "abortacabras" meaning "which causes the goats to abort").

co.: It is the glory of the limestone cliffs, from the montane to the lower lpine zone, no matter their orientation, growing in the company of *Ra-nonda myconi, Bupleurum angulosum, Potentilla caulescens*, etc.

.oc.: Santa Elena bridge, towards Forqueta del Gabieto (B); Faja Pelay, óriz ravine (O); La Pardina, Molino de Aso (A); Revilla, Los Mallos bridge E); Cirque of Pineta, trail to Añisclo Pass (P).

104 **Viola biflora** L. (Violaceae)

Twoflower Violet. Pensée à deux fleurs. Pensamiento amarillo.

SECT.: B O V A E P C. ALT.: (1325)1500 - 2690 m. H. Bor.-alp. C

'his small pansy has kidney-shaped leaves and produces yellow lowers usually in pairs, as its Latin name indicates.

co.: It tucks into shady fissures of wet, seeping rocks as well as at he foot of crags with prolonged snow cover, no matter the nature of he substratum, where it lives in the company of *Cystopteris fragilis* or *lypericum nummularium*.

.oc.: Lake Bernatuara, Batanes ravine (B); Faja Pelay, Aguas Tuertas at arriata (O); La Pardina ravine, Liana Mala (A); Foratarruego, Pleta de los 'eixins (E); Embudo de Pineta, Faja Tormosa (P).

SCREES

Screes are formed as a consequence of the wearing down of ridges and cliffs. Wide temperature fluctuations taking place between day and night provoke the frost/thaw phenomena that manage to break the rocks. These places set harsh conditions for plant life, such as the sliding movement of the substratum and the falling of stones from the adjacent cliffs, which break the plants.

All this acts as an evolutionary trigger so accordingly a number of species are endemic to this type of environment, as it also occurs on rocky ground.

105 Arabis alpina L. (Brassicaceae)

Alpine Rock-Cress. Arabette des Alpes. *Carchesia*

SECT.: B O T A E P C. ALT.: (990)1435 - 3165 m. Ch. Latebor.-alp. F

This is a rather hairy herb with pale green, toothed leaves, those of the rosette having a stalk while the others embrace the stem, which can grow up to 40 cm. Its white flowers are fairly large. This species takes on a wide variety of shapes, ranging from the slender to the robust.

Eco.: It lives on scree slopes, chiefly of limestone, from the montane to the subnival zone.

Loc.: It has its uppermost location in Spain, 3,165 m, at Pico Oriental de la Cascada. San Nicolás de Bujaruelo (B); Brecha de Rolando, Cotatuero, Mondarruego, Pico de la Cascada (O); Capradiza ravine (A); Añisclo Pass, Cirque of Pineta (P); Mountain Refuge of La Estiva (C).

106 Aquilegia pyrenaica DC. subsp. pyrenaica (Ranunculaceae)

Pyrenean Columbine. Ancolie des Pyrénées. Aguileña del Pirineo. *Palometas de puerto*

A. aragonensis Willk.

SECT.: B O A P. ALT.: (1040)1550 - 2400 m. H. Pyr.-Cant. E

The name comes from the hooked spurs of its petals, shaped like the talons of an eagle. However, this species differs from the common columbine, also present in the Pak, that it bears large flowers compared to the size of its leaves and also its straight or slightly curved spurs, never strongly hooked. It has been used as an aphrodisiac and its seeds, soaked in olive oil, were applied to battle head lice.

Eco.: This plant is an endemism to the Pyrenees and the Cantabrian Mountains that colonises scree, rock crags and stony pastureland on calcareous soil.

Loc.: Col d'Oulettes, Lapazosa ravine (B); Cotatuero (O); Sierra Custodia, San Úrbez (A); Balcón de Pineta, track towards La Estiva (P).

107 Borderea pyrenaica Bubani & Bordère ex Miégev.

(Dioscoreaceae)

Dioscorée des Pyrénées. Borderea del Pirineo

Dioscorea pyrenaica Bubani & Bordère ex Gren.

SECT.: B O V A E P C. ALT.: (1320)1550 - 2410(2500) m. G. Pyr. C. F

This one is, along with its relative B. chouardii, the herbaceous plant with the most longevity in Europe, with a lifetime of 300 plus years, much more than some trees. Its age has been calculated by means of the scar that the yearly stem leaves on the xylopodium or yam, a woody tuber. The stems bear ieart-shaped, dull green leaves. Moreover, it is the only European ılant pollinated by ants, these being capable of distinguishing the ertile pollen from the sterile. In this plant both sexes occur in diffe-ent specimens. This genus is unique to the Pyrenees, its closest re-atives living in Africa and America, where yam is consumed as food.

:co.: It takes root on the unstable limestone screes in the subalpine and ılpine zone of the Central Pyrenees.

.oc.: This plant has some of its best populations within the Park and its ricinity. Bujaruelo Pass (B); Faja Canariellos, Cirque of Soaso (O); Sierra :ustodia, Liana Mala (A); Montaña de Sesa, Angonés ravine (E); Montaspro, :stiba de Espierba (P).

108 Campanula speciosa Pourr. subsp. speciosa

(Campanulaceae)

Showy Harebell. Campanule à belles fleurs. Campanillas

SECT.: B O V A E P. ALT.: 1000 - 1900(2100) m. H. LatePyr. E

)oubtless, the most attractive and ›eautiful harebell in our flora as its .atin name states, bearing plenty of ›ell-shaped, pale blue flowers. This ›pecies has linear-lanceolate leaves ınd can reach one metre in height.

:co.: It thrives on limestone scree and oamy slopes of the montane zone along vith *Ligusticum lucidum* and *Stipa calanagrostis*, among other species.

.oc.: Its geographical range goes beyond he Pyrenees. Lapazosa ravine, Gabieto avine (B); Senda de Cazadores, Faja :anariellos (O); Betosa ravine (A); ıngonés ravine (E); Estiba de Espierba, rack to La Larri (P).

109 **Cirsium glabrum** DC. (Asteraceae)

Cirse glabre. Cardo blanco

SECT.: O V A P. ALT.: (1300)1600 - 2440 m. G. Pyr.

A thistle with slightly fleshy, very prickly, light green leaves, and leafy stems bearing at the top yellowish-white flower heads. It is hairless, hence its name *glabrum*.

Eco.: This is a remarkable endemic to the Pyrenees that colonises scree slopes and stony pastureland, particularly on limestone, loam and flysch, in the company of other plants such as *Festuca gautieri* subsp. *scoparia* or *Crepis pygmaea*.

Loc.: Cola de Caballo, Faja Pelay (O); Sierra de la Estiba, Fuen Blanca ravine (A); Estiba de Espierba, trail to Añisclo Pass (P).

110 **Crepis pygmaea** L. subsp. **pygmaea** (Asteraceae)

Crépide naine. Falsa árnica, crepis enano

SECT.: B O V A E P C. ALT.: (1000)1350 - 3134 m. G. Alp.

This plant is firmly attached t the mobile scree by means c a long rhizome, being capabl of re-emerging after gettin buried or broken due to slop motion. Its rather roundec dull green leaves blend in wit its surroundings, especially o the schist screes. It produce some yellow flower head tinged purple, which brighten the mountainsides with colour fror the Alps right to the Sierra Nevada, passing through the Pyrenees.

Eco.: Scree slopes, mainly on limestone, although it also tolerates schis from the montane belt to the subnival. It is found next to *Bordere pyrenaica*, *Carduus carlinoides*, *Linaria alpina*, etc.

Loc.: Gabieto, Planabé ravine (B); Punta de las Escaleras, Pico Millar (O); Sierra Custodia (A); Angonés ravine, La Consusa ravine (E); Easter Astazu, Balcón de Pineta.

111 Erodium glandulosum (Cav.) Willd. (Geraniaceae)

Bec-de-grue glanduleux

E. foetidum subsp. **glandulosum** (Cav.) O. Bolòs & Vigo,
E. petraeum subsp. **glandulosum** (Cav.) Bonnier, **E. macradenum** L'Hér.

SECT.: O V A E C. ALT.: 1350 - 2200 m. Ch. Late Pyr.-Cant. F

This species has much divided leaves, sticky and that grow close to the ground. But certainly its most distinctive features are the two different patterns within its corolla, with the two upper petals having a marked purple venation whilst the three lower ones are more modestly-coloured. The name of the genus comes from Greek, meaning heron, as the shape of the fruit recalls the beak of these birds.

Eco.: It is found on stony, frost-disturbed ground, as well as on ledges, screes, wind-swept ridges and so on, in the montane and subalpine zone.

Loc.: It is exclusive to the Pyrenees, the Cantabrian Mountains and the Iberic System. Cirque of Soaso, Millaris Pass (O); track to Cuello Arenas, Sestrales (A); Castillo Mayor (E).

112 Gymnocarpium robertianum (Hoffm.) Newman (Aspidiaceae)

Limestone Oak Fern, Scented Oak Fern. Polypode de Robert

Dryopteris robertiana (Hoffm.) C. Chr.

SECT.: B O A E P. ALT.: 1030 - 2560 m. G. Bor.-alp. E

This fern has a slender creeping rhizome and triangular blades, with small glands. The species was given its name because of its odour resembling that of *Geranium robertianum*.

There is another similar species in the Park, ***G. dryopteris*** (small image), the Western Oak fern, Common Oak Fern or Northern Oak Fern, that is characterised by the absence of glands and grows in woodland, chiefly on acidified humus.

Eco.: This species colonises stony ground and limestone screes, from the montane to the alpine zone together with plants such as *Valeriana montana* or *Rumex scutatus*.

Loc.: Ordiso (B); Turieto, Faja Pelay (O); Cuello Arenas, Añisclo gorge (A); Foratarruego (E); Faja Tormosa, Montaspro (P).

113 Ligusticum lucidum Mill. subsp. lucidum

(Umbelliferae)
Ligustique des Pyrénées. Ligústico
L. pyrenaeum Gouan
SECT.: B O T E P. ALT.: 1000 - 2070 m. H. Latesubmed. **F**

Perennial, with much divided leaves and white flowers. The epithet *lucidum* means brilliant, because of the shine on its leaves, whilst the name of the genus refers to the Italian region of Liguria, where it is also found.

Eco.: It thrives on sunny screes of the montane and lower subalpine zone, sometimes even on river gravel or other stony ground, in the company of *Stipa calamagrostis* or *Campanula speciosa*.

Loc.: Las Comas ravine, Ordiso (B); Faja Racón, Gradas de Soaso (O); Castillo Mayor (E); La Larri, Montaspro (P).

114 Ranunculus parnassifolius L.

subsp. **heterocarpus** Küpfer (Ranunculaceae)
Parnassus Buttercup. Renoncule à feuilles de parnassie.
Ranúnculo de glera
SECT.: B O V A E P C. ALT.: (1550)1700 - 2905 m. H. Alp. **C**

This buttercup has dull green leaves that blend in very well with the environment and which, because of its oval shape, resemble those of the Grass-of-Parnassus (*Parnassia palustris*), hence its name. This subspecies in particular is distinguished from others occurring in the Pyrenees or the Cantabrian Mountains either by having a smaller number of stamens or by the fact that many achenes are aborted. Like many buttercups, it is a toxic plant that has been used in folk medicine as well.

Eco.: Its peculiar root system allows this plant to colonise mobile limestone screes from the subalpine to the subnival belt, together with *Borderea pyrenaica* or *Veronica nummnularia*.

Loc.: Slopes of Monte Perdido, Descargador Pass (O); Cuello Gordo, Sierra de la Estiba (A); La Valle, Montaña de Sesa (E); Faja Tormosa, Balcón de Pineta (P).

115 **Stipa calamagrostis** (L.) Wahlenb. (Poaceae)

Calamagrostide argentée
Achnatherum calamagrostis (L.) Beauv., **Calamagrostis argentea** (Lam.) DC.

SECT.: B O T A E P C. ALT.: 730 - 1900 m. H. Submed. F

'his relatively tall grass is easily recognisable because of its long ;leaming inflorescence. It has a powerful root system that allows it o produce new stems after soil motion or breakage.

:co.: It colonises limestone screes and slopes in the montane zone, usua-ly on sunny spots, together with *Rumex scutatus* or *Galeopsis ladanum*.

.oc.: Santa Elena bridge, San Bastán ravine (B); Andecastieto, Faja Racón O); San Úrbez, Betosa avine (A); Yaga gorge, ıngonés ravine (E); El 'elqueral, Estiba de Es-ɔierba (P).

116 **Veronica nummularia** Gouan

subsp. **nummularia** (Scrophulariaceae)
Véronique nummulaire. Verónica de glera

SECT.: B O A E P C. ALT.: (1680)1900 - 3100 m. Ch. Pyr. C

A creeping plant, dark green in colour and bearing blue flowers. It is little conspicuous because of its appearance. As the name of this species indicates, its rounded leaves look like coins.

Eco.: This endemic plant to the Pyrenees thrives on high mountain screes and limestone ridges, from the subalpine to the subnival zone, sharing the same habitat as *Crepis pygmaea* or *Ranunculus parnassifolius* subsp. *heterocarpus*.

Loc.: Picos de la Cascada, Monte Perdido (O); Cuello Arenas, Morrón de Arrablo (A); Tres Marías, Sources of the Yaga (E); Pico Astazu, Embudo de Pineta (P).

HIGH MOUNTAIN

Above 3,000 metres and as a consequence of prolonged snow cover and freezing soil, that shorten the vegetative period of plants, environmental conditions become extremely harsh. As compensation, the snow cover both insulates from the cold and retains heat, moreover supplying water, essential for life, after thawing. Wind sweeps the snow off of the ridges and erodes and dries the soil. Strong temperature variations cause rock breakage, cryoturbation and landslides. Powerful ultraviolet radiation can cause mutations while pollination through insects is very restricted, its activity being reduced or null because of cold and strong winds. However, a handful of plants is capable of enduring the harsh high mountain climate. As for the Pyrenees, from amongst the more than 3,500 species living there, only around 150 can achieve this feat; more specifically, 95 of them are found on the 34 summits above 3,000 metres of Ordesa and Monte Perdido National Park and its peripheral zone.

High mountains are not the appropriate place for opportunistic plants, those with a short or annual life-cycle, putting all of their eggs quite literally in one basket. Accordingly, their first strategies to survive are either longevity or vegetative propagation. Bright-coloured flowers allow them to attract the few pollinating insects that can reach such elevations, and this is the case of purple colour in *Linaria alpina* subsp. *alpina* and *Saxifraga oppositifolia*, or the pink in *Silene acaulis*. If insects are so scarce that the aforementioned *Saxifraga* can not be pollinated, then it is capable of self-pollinating.

Most of these plants fight cold and drought by means of cushion-like structures that reduce the surface area exposed to heat and humidity loss, as it occurs with *Silene acaulis* or *Minuartia sedoides*. Also, other species such as *Saxifraga pubescens* or *Androsace ciliata* possess white hairs that allow them to retain warm air and humidity and also serve to protect themselves against ultraviolet radiation, as is the case with *Cerastium alpinum* or *Leucanthemopsis alpina*. Soil motion is minimized by having long, deep roots.

117 **Androsace ciliata** DC. (Primulaceae)

Androsace ciliée. Androsace ciliada

SECT.: B O A E P C. ALT.: (2180)2400 - 3330 m. Ch. Pyr. F

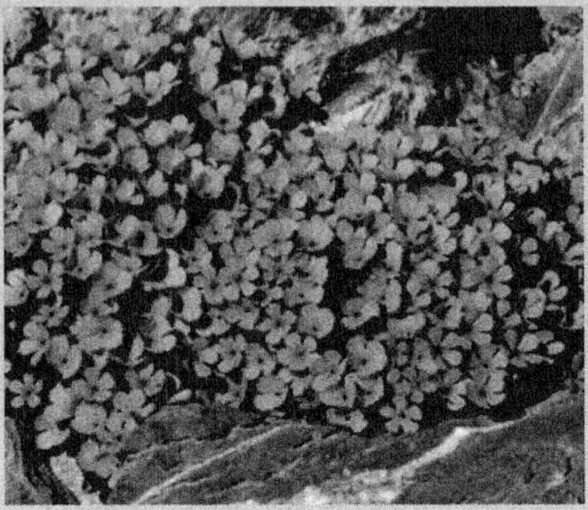

A gorgeous plant endemic to the Pyrenees forming dome-shaped cushions decked out with pink flowers. Its name refers to the fine hairs (cilia) on the tip of its leaves.

Eco.: It inhabits ridges, summits and rather stable screes on the highest peaks (alpine and subnival belts), no matter their geological nature, together with *Saxifraga pubescens* or *Minuartia cerastiifolia* in these extreme environments.

Loc.: Headwaters of the Ara, Bernatuara Peak (B); Casco, Faja de las Flores; Mondarruego, Punta de las Escaleras (O); Tres Marías (E); Lake Marboré, La Munia lakes (P).

118 **Artemisia umbelliformis** L. (Asteraceae)

White Genepi. Genépi blanc. Artemisa de montaña

SECT.: B O A E P C. ALT.: (2180)2400 - 3330 m. Ch. Alp. E

Cushion-like perennial, densely covered with white or greyish silky hairs. The yellow flower heads are usually grouped to form an umbel, hence its surname. This species belongs to the alpine genepis used in the distillation of digestive liqueurs such as «Chartreuse» or «Benedictine».

Eco.: It grows on initial cryoturbated soils of the fissures and ledges of the Park summits, being indifferent to the substrate.

Loc.: Headwaters of the Ara, Bernatuara Peak (B); Casco, Faja de las Flores; Mondarruego, Punta de las Escaleras (O); Tres Marías (E); Lake Marboré, La Munia lakes (P).

119 **Cerastium alpinum** L. (Caryophyllaceae)

Alpine Mouse-Ear Chickweed, Alpine Chickweed. Céraiste des Alpes. Melosilla, oreja de ratón

SECT.: B O A E P C. ALT.: 1740 - 3210 m. Ch. Bor.-alp. E

In this genus the seeds are kept inside a fruit called a capsule, crowned with little horns or teeth, hence the name Cerastium, which comes from Greek, meaning horned. This species is a perennial forb whose hairiness may vary to a certain extent, being sometimes lanate (woolly).

Eco.: It lives on the screes and wind-blasted ridges of the high mountains, together with *Arenaria moehringioides* or *Kobresia myosuroides*.

Loc.: Bernatuara summer pastureland, Bujaruelo Pass (B); Gabietos, Mondarruego (O); Punta Navarro, Fuen Blanca ravine (A); Tres Marías (E); Robiñera, La Munia (P).

120 **Draba dubia** Suter subsp. **laevipes** (DC.) Braun-Blanq. (Brassicaceae)

Drave à pédicelle glabre

D. laevipes DC., **D. tomentosa** Clairv. var. **frigida** (Saut.) Gren & Godr.

SECT.: B O V A E P C. ALT.: 1650 - 3200 m. Ch. Med. Mount.-Alp. F

Small perennial forming several rosettes in turf and a proportionally big silicula (the fruit). The epithet *laevipes* means smooth, referring to the base of the fruit, which is hairless.

Eco.: It colonises rock fissures, preferably on limestone but sometimes also on schists, of high mountain sunny spots, whilst it is found in the shade at low-lying locations.

Loc.: CCol d'Arratille, Plana de Alba Pass (B); Brecha de Rolando, Lago Helado, Mondarruego, Senda de Cazadores (O); Sestrales (A); Balcón de Pineta (P); La Valle (E); Montaña de Ruego (C).

Some other species of *Draba* are also found within the Park: thus, in the higher mountains live ***D. aizoides*** subsp. ***aizoides*** (yellow-flowered, small image), or *D. siliquosa* subsp. *carinthiaca* and D. *tomentosa* subsp. *ciliigera*, the two latter especially at the siliceous peripheral zone. Finally, ***D. hispanica*** subsp. ***hispanica***, reaches its northernmost distribution limit on some rock crags of the subalpine belt at Ordesa (small image in bottom right corner).

121 Linaria alpina (L.) Mill. subsp. alpina

(Scrophulariaceae)

Alpine Toadflax. Linaire des Alpes. Violeta de glera

SECT.: B O T V A E P C. ALT.: (1300)1500 - 3300 m. Ch. Alp. C

Small creeping herb with fleshy, glaucous-coloured linear leave which inhabits stony high mountain slopes. It is particularly evider thanks to its violet flowers tinged with orange hues.

Eco.: It is found on screes and ridges of different kind of rocks, reaching th summits of the Pyrenees along with *Androsace ciliata* or *Pritzelago alpina.*

Loc.: Otal, Batanes ravine (B); Picos de la Cascada, Mondarruego (O Southern Punta Custodia, Punta de las Loseras (A); Montaña de Sesa (E Lake Marboré, Eastern Astazu (P).

122 Minuartia cerastiifolia (Ramond ex DC.) Graebn

(Caryophyllaceae)

Minuartie à feuilles de céraiste

Arenaria cerastiifolia Ramond ex DC.

SECT.: B O A E P. ALT.: 2200 - 3300 m. Ch. Pyr. C-W.

Another hairy creeping forb, the only one of its genus with ova lanceolate leaves resembling those of a *Cerastium*, hence its Lati name. This genus is dedicated to Juan Minuart, a pharmacist an botanist from Barcelona, who lived in the 18th century. It was Ra mond, the naturalist par excellence of the Monte Perdido massi who first described this plant from the Cirque of Troumouse, on th other side of Pineta, where it is abundant in the Marboré area. W can also find in siliceous sandstone outcrops ***M. sedoides*** (L.) Hier (small image).

Eco.: It is an endemic to the central and western Pyrenees, from Benasque to the Roncal. It colonises the limestone screes from the alpine belt up to the highest summits.

Loc.: Vignemale (B); Taillón, Monte Perdido (O); Punta de las Olas (A); Tres Marías (E); Astazu, Balcón de Pineta (P).

123 Potentilla nivalis Lapeyr. subsp. nivalis (Rosaceae)

Snowy Cinquefoil. Potentille des neiges. Potentilla de las nieves

SECT.: B O T V A E P C. ALT.: (1500)1800 - 3134 m. H. Alp. C

A cinquefoil with leaves divided in 5-7 leaflets and covered with a hairiness that gives them a pale greyish-green tint. As well as other high mountain species of plants, it has developed some reproductive mechanisms which enable them to survive in places where only few pollinators can climb. In this particular case, its flowers are barely open, its strategy being self-pollination.

Eco.: As its name indicates, it lives in the highest part of the mountains, colonising rock fissures, screes and wind-swept ridges, from the subalpine to the subnival zone.

Loc.: Plana Cuasta, Espelunz (B); Faja de las Flores, Millaris (O); Liana Mala, Punta de las Olas (A); Foratarruego, Pleta de los Faixins (E); Faja Tormosa, Western Astazu (P).

124 Pritzelago alpina (L.) Kuntze subsp. alpina (Brassicaceae)

Chamois Cress. Cresson de chamois

Hutchinsia alpina (L.) R. Br., **Hornungia alpina** (L.) O. Appel

SECT.: B O A E P C. ALT.: (1000)1600 - 3320 m. Ch. Alp. F

The basal leaves of this tiny herb are divided and its small flowers, white. The genus is dedicated to the Polish botanist George Pritzel, noted for his work on the *Thesaurus* of Botanical Literature.

Eco.: It lives hidden in shady rock crevices and also appears on screes and snow-cloaked stony pastureland, mainly on limestone substrate, on the highest peaks of our mountain range.

Loc.: It has been found near the summit of Monte Perdido, its uppermost location in the Pyrenees. Nevertheless, there are some abyssal populations of this plant at Ordesa, Añisclo and Pineta. Lapazosa ravine, Vignemale (B); Lago Helado, Cilindro de Marboré (O); Punta de las Olas, La Fuen Blanca ravine (A); Tres Marías (E); Eastern Astazu, Balcón de Pineta (P).

125 Salix retusa L. (Salicaceae)

Saule à feuilles émoussées. Sauce enano rastrero

SECT.: B O A E P. ALT.: 1800 - 3015 m. Ch. Alp. E

This willow is a ground-hugging shrub, forming small dense green patches which stand out among the stony harsh environment, unpropitious to vascular plants. The name of the species, *retusa*, refers to the shape of the leaves, truncate or slightly emarginate (with a notch at the apex).

Eco.: It grows underneath snow beds, chiefly on limestone substrate, as a part of dense subalpine and alpine pastureland, at times together with other dwarf willows such as the Least Willow or Snowbed Willow, *S. herbacea* (small right image) or the Net-leaved Willow, *S. reticulata* (small left image), scarcer as they prefer acidified soils.

Loc.: Escuzana, Bujaruelo pass (B); Monte Perdido, Tobacor (O); collado de Añisclo (A); Balcón de Pineta, ibones de La Munia (P).

126 Saxifraga oppositifolia L. (Saxifragaceae)

Purple Saxifrage. Saxifrage à feuilles opposées. Rompepiedras de hojas opuestas

SECT.: B O T V A E P C. ALT.: (1100)1600 - 3320 m. Ch. Bor.-alp. C

It is a creeping perennial forming a loose mat, with opposite curved little leaves, hence its specific epithet. It stands out among other saxifrages for its pink or purple flowers.

Eco.: It thrives on screes, wind-swept ridges, stony pastureland and snow beds, together with *Saxifraga pubescens* or *Androsace ciliata.*

Loc.: It is one of the species that can reach higher elevations in the Pyrenees. However, on the contrary, it is found at only 1,100 m in Turieto Bajo (Ordesa), its lowermost population in our range, in a place subject to temperature inversion. Ordiso, Lapazosa ravine (B); Gabietos; Monte Perdido (O); Sierra Custodia, Morrón de Arrablo (A); Montaña de Sesa, Tres Marías (E); Eastern Astazu, Cinca waterfalls (P).

127 **Saxifraga pubescens** Pourr. (Saxifragaceae)

Saxifrage pubescente. Rompepiedras pelosa
S. iratiana F.W. Schultz

SECT.: B O A E P C. ALT.: 2100 - 3300 m. Ch. Pyr. **E**

\ cushion-forming saxifrage with slender stems covered with palm-ike divided leaves, glandulous-hairy as its Latin name *pubescens* mplies, and white flowers. It is an endemism to the Pyrenees, with wo subspecies: *iratiana*, with a central-western distribution and no articular preferences as for the nature of the soil; and *pubescens*, rom the central and eastern Pyrenees, prone to grow on siliceous ock, although in the Park area both subspecies overlap and seem to xchange biological traits and ecology.

Growing at these high elevations we can also find another species of the same genus, this one with greenish-yellow flowers, the Musk axifrage, ***S. moschata*** Wulfen (small image).

Eco.: It is one of the few plants capable of colonising, together with *Androsace ciliata* or *Pritzelago alpina*, the barren summits above 3,000 metres, no matter the nature of the rocks.

Loc.: Bujaruelo Pass, Batanes (B); Taillón, Torre de Marboré (O); Punta de las Olas (A); Tres Marías (E); Eastern Astazu, La Munia (P).

128 **Silene acaulis** (L.) Jacq. (Caryophyllaceae)

Moss Campion, Cushion Pink. Silène sans tiges.
Musgo florido

SECT.: B O T V A E P C. ALT.: 1650 - 3250 m. Ch. Bor.-alp. **C**

This is another cushion-forming plant to be found in high mountain areas, composed of a mass of tightly-packed miniature stems with tiny leaves, giving it he appearance of a flowering moss, as referred to by some of its vernacular names, which enables it to defend itself against the erosive agents and desiccation. In early summer this plant is bedecked with dozens of iny pink flowers almost stalkless covering the entire plant.

Eco.: It inhabits screes and ridges on cryoturbated, long-lasting snow-covered ground of our highest and emblematic peaks. Other species such as *Saxifraga moschata* or *Cerastium alpinum* are among its faithful companions.

Loc.: Espelunz, Bernatuara summer pastureland (B); Lago Helado, Faja de as Flores (O); Sestrales, Cuello Gordo (A); Tozal de San Vicenda, Angonés avine (E); Tucarroya, Balcón de Pineta (P).

Bibliography

ALDEZÁBAL, A. (2001). *El sistema de pastoreo del Parque Nacional de Ordesa y Monte Perdido (Pirineo Central, Aragón). Interacción entre la vegetación supraforestal y los grandes herbívoros.* 317 pp. Investigación. Consejo de Protección de la Naturaleza de Aragón. Zaragoza.

AIZPURU, I., C. ASEGINOLAZA & al. (1999). *Claves ilustradas de la Flora del País Vasco y territorios limítrofes.* 831 pp. Servicio Central de Publicaciones del Gobierno Vasco. Vitoria-Gasteiz.

BENITO ALONSO, J.L. (2012). *Catálogo florístico del Parque Nacional de Ordesa y Monte Perdido (Pirineo aragonés).* Segunda edición corregida. Monografías de Botánica Ibérica, nº 5. Edita Jolube Consultor Botánico y Editor. Jaca (Huesca).

BENITO ALONSO, J.L. (2012). *La vegetación del Parque Nacional de Ordesa y Monte Perdido (Pirineo central aragonés).* Segunda edición corregida. Monografías de Botánica Ibérica, nº 6. Edita Jolube Consultor Botánico y Editor. Jaca (Huesca).

BOLÒS, O., J. VIGO, R. MASALLES & J.M. NINOT (2005). *Flora manual dels Països Catalans.* 1247 pp. 3ª edición. Editorial Pòrtic. Barcelona.

CASTROVIEJO, S., coord. (1986-2014). *Flora Iberica.* Real Jardín Botánico, CSIC. Madrid.

GÓMEZ, D. -ed.- (2012). *Atlas de la flora vascular de Aragón.* Instituto Pirenaico de Ecología, CSIC y Gobierno de Aragón. Recurso electrónico en proyectos.ipe.csic.es/floragon/.

LÓPEZ GONZÁLEZ, G. (2004). *Guía de los árboles y arbustos de la Península Ibérica y Baleares. (Especies silvestres y las cultivadas más comunes).* Mundi Prensa Libros, S.A. Madrid.

NUET BADIA, J. (2008). *Plantes alpines dels Pirineus.* 190 pp. Publicacions de la Abadia de Montserrat, Col. Cavall Bernat, 59. Barcelona.

MAZA, M., F. CARTAGENA & L.M. NAVARRO (2005). *Guía de flores del Pirineo.* 408 pp. Barrabés Editorial.

MUÑOZ, C. (2014). *Orquídeas de Aragón.* 202 pp. Col. Guías Imprescindibles de flora, nº 2. Edita Jolube Consultor Botánico y Editor. Jaca (Huesca).

PUENTE, J. & J.L. BENITO (2014). *Guía imprescindible de las flores del Prepirineo y tierras vecinas.* 204 pp. Col. Guías Imprescindibles de flora, nº 3. Edita Jolube Consultor Botánico y Editor. Jaca (Huesca).

SAULE, M. (1991). *La grande flore illustrée des Pyrénées.* 765 pp. Editorial Milán, Toulouse & Tarbes.

URIBE-ECHEBARRÍA, P. M. & I. ZORRAKÍN (2004). *Claves ilustradas de la Flora del Moncayo.* 335 pp. Departamento de Medio Ambiente del Gobierno de Aragón. Zaragoza.

VIDALLER, R. (2004). *Guía de árboles y arbustos del Pirineo aragonés.* 358 pp. Barrabés Editorial. Cuarte (Huesca).

VIGO, J. (2008). *L'Alta muntanya catalana: flora i vegetació.* 2ª edición. 443 p. Institut d'Estudis Catalans, Secció de Ciències Biològiques y Centre Excursionista de Catalunya. Barcelona.

VILLAR, L. & J.L. BENITO ALONSO (2001). *Memoria del mapa de vegetación actual del Parque Nacional de Ordesa y Monte Perdido,* escala 1: 25 000. 145 pp. (incluye mapa 1:25 000 en tres hojas). Serie Técnica. Organismo Autónomo Parques Nacionales. Ministerio de Medio Ambiente. Madrid.

VILLAR, L., J.Mª. PALACÍN, C. CALVO, D. GÓMEZ & G. MONTSERRAT (1992). *Plantas Medicinales del Pirineo Aragonés y demás tierras oscenses.* 2ª edición. Diputación Provincial de Huesca.

VILLAR, L., J.A. SESÉ & J.V. FERRÁNDEZ (1997-2001). *Atlas de la Flora del Pirineo Aragonés, I y II.* 1551 pp. Consejo de Protección de la Naturaleza de Aragón e Instituto de Estudios Altoaragoneses. Huesca.

Cartography

Mapa del Parque Nacional de Ordesa y Monte Perdido a escala 1: 25.000. Serie Parques nacionales de España, nº 1. Editan los Ministerios de Medio Ambiente y Fomento y PRAMES. Madrid, 2000.

Alphabetical index of scientific and vernacular names

The scientific names in bold type correspond to the species having a plant profile or an image, while the synonyms are in italics. Vernacular and common names in English, French or Spanish are in Roman print whilst the Aragonese are in italics.

D

E

F

G

H

I

J

K

L

T

U

V

W

X

Y

Z

Printed in Dunstable, United Kingdom

91342770R10058